Construction Collaboration Technologies

The extranet evolution

Paul Wilkinson

D1347695

Taylor & Francis
Taylor & Francis Group

LONDON AND NEW YORK

First published 2005
by Taylor & Francis
2 Park Square, Milton Park, Abingdon, Oxon OX14 4RN

Simultaneously published in the USA and Canada
by Taylor & Francis
270 Madison Ave, New York, NY 10016

Taylor & Francis is an imprint of the Taylor & Francis Group

Typeset in Sabon by
Newgen Imaging Systems (P) Ltd, Chennai, India
Printed and bound in Great Britain by
TJ International Ltd, Padstow, Cornwall

British Library Cataloguing in Publication Data
A catalogue record for this book is available from the British Library

Library of Congress Cataloging in Publication Data
Wilkinson, Paul.
    Construction collaboration technologies : the extranet evolution /
Paul Wilkinson.
      p. cm.
    Includes bibliographical references and index.
    1. Extranets (Computer networks). 2. Interorganizational relations.
3. Building. I. Title.
TK5105.875.E87W548 2005
624'.068–dc22   2005005407

ISBN 0–415–35858–2 (hbk)
ISBN 0–415–35859–0 (pbk)

Contents

Figures

Tables

Foreword

The UK construction industry has come a long way since my report, 'Constructing the Team', was published in 1994 and the subsequent Egan process since 1998. Many of these reports' recommendations have been implemented, and numerous client, contractor, consultant and supplier organisations have begun to adopt more integrated, collaborative styles of working. This is now also a headline target for the Strategic Forum for Construction. However, there is still a long way to go.

This gradual but, I believe, irreversible shift towards partnering-type approaches to project delivery contrasts greatly with the enthusiastic adoption of mobile working and internet technologies within parts of the industry. In 1994, only about one in a hundred construction businesses had an internet connection. Today, many of us now use email, websites, laptops and mobile telephones almost without a second thought.

Internet technologies have been effectively employed to support collaborative working. Over the past five years, construction collaboration technologies have moved from being an industry novelty towards almost routine deployment by some organisations on medium-to-large projects. But this market is still maturing and has not yet reached the stage when most professionals will have sufficient knowledge to make informed decisions. Anyone working on, or aspiring to work on, major projects will increasingly need to be familiar with these systems and the pros and cons of their use. There is still much to do to involve specialist and trade sub-contractors in this whole approach.

I first became aware of construction collaboration technologies in the late 1990s. In 2000, I agreed to become deputy chairman of BIW Technologies and met the company's head of corporate communications, Paul Wilkinson. I have since seen Paul expand his role to become involved with the sector's own trade association, the Network of Construction Collaboration Technology Providers. He also set himself a major challenge: to write the industry's first comprehensive and authoritative guide to the technologies.

Paul has risen to this challenge very well. This is a timely and stimulating book, aimed at industry participants wanting to know more about the construction collaboration technology market in the United Kingdom. Industry customers, their advisors, professionals from technician level practitioners at the project work-face up to board-level decision-makers, and students looking to enter the industry, will all find this book enormously informative and interesting. It answers many of the questions commonly asked about the technologies. It includes real-life examples to illustrate its key points, features several case studies, and has a useful glossary.

Paul has drawn on his insider's knowledge and experience and those of many others, both in the United Kingdom and in other markets, ranging from industry practitioners to academics. Chapters in this book describe the emergence of the technologies and the early development of the UK market. He gives a detailed description of the principal generic features of the various systems, debates the pros and cons of externally hosting and looks at connectivity issues. However, the book is certainly not IT-focused and will be a valuable help to the non-technically minded.

Paul has expanded on his previously published work on the legal issues associated with collaboration technologies. His chapter on human issues influencing technology adoption is particularly apposite. It emphasises that successful deployment of such systems is much more about people and processes than managing 'bits and bytes'. He assesses evidence of the benefits of using the systems, and he concludes by looking at what the future might hold.

As the twenty-first century proceeds, we will certainly see further integration of IT into working practices in the UK construction industry. It will become part of our normal everyday routine. In the meantime, clients need to insist on collaborative working in order to deliver their projects on a team-working basis and in a non-confrontational manner which strips out non-value-adding activities. By using IT to support more progressive approaches, we can encourage an increasingly large proportion of the industry to change for the better. The ultimate beneficiaries will be clients and the country as a whole.

Sir Michael Latham
January 2005

Preface

Why write this book?

As the twentieth century drew to a close, investors, company executives, entrepreneurs and journalists were all hugely enthusiastic and optimistic about the potential of the internet. There was talk of a 'dot.com boom', and the architecture, engineering and construction (AEC) market was not immune from this obsession. From the late 1990s until about mid-2000, the UK construction-related trade press published numerous e-commerce stories – towards the end, it seemed that hardly a month passed without a new construction portal or e-marketplace being launched.

In 2000, however, the dot.com bubble burst, and many of the ambitious new AEC e-businesses quickly disappeared. But not all of them. Interest in a subset of e-commerce technologies – internet-based collaboration (or 'c-commerce') systems – began to grow. IT analysts such as the Gartner Group, business process re-engineering gurus including Michael Hammer (2001) and James Champy (2002), and major software vendors such as Microsoft all felt this was a sector due for significant growth. Within the AEC industry, collaboration technologies had begun to attract some overdue attention, first in the United States, and then in other countries. In the UK market, a handful of new start-ups, plus some existing technology businesses, began to offer internet-based software applications that allowed construction project teams to share drawings and documents electronically.

With the internet increasingly an established part of many construction businesses' communications, many professionals were interested in using such collaboration technologies, but were – quite understandably, following the hype surrounding web-based trading exchanges and other ill-starred ventures – nervous about taking the plunge. It was, perhaps, easy to justify investing time and money in more mainstream, locally managed AEC applications such as computer-aided design (CAD), project management and cost estimating, which improved individual productivity. But it was less easy to justify investments in new types of applications, from often previously unknown technology businesses, that, in many instances, were not even hosted by any member of the project team.

It was – and to some extent still is – a substantial market education challenge. Vendors such as 4Projects, BIW and Business Collaborator had to build a generic awareness of construction collaboration technologies and convince industry clients and their project teams that they offered significant advantages over existing communication methods (a task made slightly easier given the increased stress placed on 'partnering' or 'collaborative working' over the 10 years or so since the Latham Report). Only then could they start to respond to the more detailed questions typically asked by prospective customers and end-users seeking to differentiate between the competing vendors and their systems. By early 2005, the market education process was continuing; according to one survey (IT Construction Forum 2004), more than half of all projects were still managed using more conventional communication channels such as email, fax and post.

Helping to fill the information gap and to push the collaborative working message, several UK trade associations and membership organisations (e.g. the Construction Industry Computing Association, Construct IT, the Institution of Civil Engineers and the Royal Institution of Chartered Surveyors, to name a handful) have produced reports, guides and case studies offering an introduction to, or independent guidance on, the technologies and the principal providers. Some of these documents are also available online, and – as one might expect for web-based technologies – several websites offer further information and support (e.g. Emap's Construction Plus guide to Project Extranets, the IT Construction Forum, and – launched in December 2003 – the Network of Construction Collaboration Technology Providers). A steady stream of conferences, seminars and exhibitions are also either wholly or partly devoted to the selection, utilisation and benefits of the technologies.

These guides, reports and case studies, websites, and conferences and other events – augmented by the flashy websites, glossy marketing literature and glib product demonstrations of the vendors themselves – provide prospective and existing customers and users with lots of useful information (and some, unfortunately, that is less useful). However, it is an onerous task to draw all this information together, let alone sort, interpret and analyse it so that one can make informed and fully reasoned decisions. This is what prompted this book.

This book draws together much of the available relevant information to provide a comprehensive and independent guide to the technologies. Drawing on some ten years' experience of researching and writing about the construction industry and IT, including five spent as part of the team at BIW Technologies, I aim to help answer some of the most frequently asked questions relating to construction collaboration technologies. Each of the following chapters is devoted to discussion of a key question (Who hosts the systems? How do we connect to them? What are the legal issues? What are the benefits? etc.). These questions are outlined at the end of the first chapter; depending on their needs, readers can either dip into the relevant chapters or read the whole book to get a complete overview of the sector. To further help the reader, each chapter starts with a few outlining bullet points, and the key issues are summarised at its conclusion. A glossary is also included to explain some of the technical terms, and the bibliography provides numerous references for readers wanting to examine subjects in greater detail.

As a whole, this book concentrates mainly on the AEC industry in the United Kingdom. I make no apologies for this. I expect most readers will want to learn and understand more about the technologies available in and appropriate to their home market. There would be little point in me trying to produce a detailed global account when few UK-based technology vendors yet have truly international operations (at least with respect to their collaboration offerings), and most vendors in other countries have similarly focused on their own home markets. I have therefore concentrated on describing the development, providers, infrastructure, legal issues, etc., that relate to the UK AEC industry. But where relevant, this book also draws on information from other markets – mainly, but not exclusively, from the United States (incidentally, there must surely be scope for a similar book covering the American AEC market). And some issues – for example, choosing a vendor, technology features, hosting, connectivity, people issues, benefits, etc. – are quite generic, and non-UK readers will hopefully find much information of interest and value to them.

Paul Wilkinson
January 2005

Acknowledgements

Numerous friends, colleagues and acquaintances have contributed to my understanding of the UK construction collaboration technologies sector over the past few years.

I would particularly like to thank all my colleagues at BIW. In particular BIW chief executive Colin Smith has supported this project from the beginning, helping me develop key ideas and, drawing on 20 years experience in construction IT, providing invaluable feedback on early drafts of the text. I am also grateful to BIW deputy chairman Sir Michael Latham for writing the foreword. In my frequent dealings with them, Andrew Boaden, Bill Flind, Steve Cooper, Narinder Sangha, Brandon Parkes, Suzie Ballard, Guy Hussey, Danny Polaine, Duncan Kneller, Asif Sharif, Lisa Gledhill, Ed Boxall, Chris Stebbings, Richard Parker, John Osborne, Aneel Khanna, Matthew Ottaway, Nick Sansome, Simon Price and George Stevenson have all been very helpful, as have the guys from BIW's Nottingham office, notably Chris Aldridge, Mitul Sudra, Shailet and Mita Patel, and Adam Sliwinski.

Many customers and end-users have also been influential, including Jasper Singh and Peter Dampier at Gleeds, Crawford Patterson and Danyal Kola at Mace, Seamus Mockler at Kajima, Steve Jones from KMI Water, Glyn Hughes of United Utilities, Tony Brown at Bucknall Austin, Steve Smith, formerly of Sainsbury's, Stephen Boid from Crest Nicholson and James Ballantyne at Interserve.

As well as BIW, several of the other construction collaboration technology vendors have also contributed to this book. I would like to thank: Duncan Mactear of 4Projects, Sanjeev Shah and Steve Crompton at Business Collaborator, Francis Newman at Cadweb, Franco Iannaccone of Causeway, and Graham Howarth and Jeremy Sainter at Sarcophagus. Special thanks to 4Projects, BIW Technologies, Business Collaborator, Cadweb, Causeway and Sarcophagus for permission to reproduce the screenshots shown in Chapter 5. And thanks also to Bill Healy and Tim Broyd at CIRIA, managers of the Network of Construction Collaboration Technology Providers (NCCTP), the umbrella group for the above mentioned organisations.

I am also particularly indebted to Ming Sun at the University of West of England and Andrew Magub of Artas Architects in Australia who also cast impartial eyes over early drafts of the text. Thanks also to: Peter Goodwin, Andrew Scoones and John George of the PIX Protocol team at the Building Centre Trust, Ross Sturley and Seb Byrd of Emap, Mike Davis at the Butler Group, Anna McCrea of the IT Construction Forum, Leslie and Mark Mitchell at Pattison Mitchell, Chris Preece at University of Leeds, Cathrine Ripley at FSP Law, David Savage of Hammonds Solicitors, Bhzad Sidawi at Cardiff University, Don Ward and Richard Saxon at Be, and David Whitton at Wren Insurance. I also had some invaluable exchanges with researchers in the United States, including Burcin Becerik at Harvard University, 'Frank' Pollaphat Nitithamyong at Purdue University and the magisterial Joel Orr of Cyon Research.

I have also undoubtedly benefited from countless conversations with individuals at industry events across the United Kingdom, and I acknowledge that the roots of some of my arguments may well lie in some of those discussions. However, having condensed these perspectives and masses of other data into the following ten chapters, any errors or omissions are, of course, entirely my responsibility.

Last but not least, I must thank my wife Helen and two children Frankie and Gus for providing love, support and sometimes much-needed diversion throughout the writing of this book.

Paul Wilkinson
January 2005

Chapter 1

Defining collaboration

This chapter:

- briefly describes the development of Internet-based collaborative applications specific to the UK architecture, engineering and construction (AEC) industry;
- refines an AEC-specific definition of 'construction collaboration technologies';
- outlines the subjects covered in the remainder of the book, helping readers to quickly find the topics that interest them.

In the early years of the twenty-first century it is increasingly difficult to imagine working without the internet.[1] Yet only a decade earlier, the World Wide Web (WWW) was still in its infancy and 'spam' was a proprietary meat-based product celebrated by the *Monty Python* television comedy team. Today, almost every sizeable organisation has its own website. And email, with its ability to support attachments from numerous other software programs, has become almost ubiquitous for day-to-day written communication; indeed, it has also become a victim of its own success – spawning a steady flow of newspaper articles about viruses, security lapses and 'email overload'.

Since the late 1990s, the explosive growth of the internet coupled with the development of better telecommunications links has also provided a platform for many new types of information technology (IT).[2] Most notably for the purposes of this book, there has been a surge in demand for applications that allow groups of people to collaborate with each other. For the highly information-dependent and cost-conscious AEC industry, where projects are routinely delivered by relatively short-lived, multi-disciplinary, multi-company, multi-location groups of people, the opportunity to use IT to send and receive large volumes of project data over longer distances more quickly and cheaply was too good to miss.[3] Around the world, software businesses recognised the opportunity and began to develop applications to capitalise on the growing AEC demand for more efficient team communication.

This book focuses on a particular group of applications sometimes described as, among other things, 'extranets'. This chapter discusses the terminology and attempts to develop a more appropriate alternative description. It proposes a term used by the leading UK vendors themselves – 'construction collaboration technologies' – and relates its definition to the overall development of collaborative working practices within the AEC industry.

This book is also focused primarily on the AEC industry in the United Kingdom. While there have been similar developments in other parts of the world, most notably in the USA but also in mainland Europe, Australasia, Canada, Israel and South Africa (to name a few), the fragmented nature of the international AEC industry, the relative immaturity of most of the software businesses involved, and their initial focus on developing products and services to serve their domestic markets, have so far largely precluded any of the vendors from marketing their products and services globally.[4] This will change, of course, as the technology becomes more widely accepted, as vendors mature, consolidate and expand, etc.; UK experience will also be influential in the adoption of the technology in other national markets, particularly those whose AEC project delivery methods follow UK models (the same also applies in reverse, of course: for example, one Australian-based vendor has opened a UK office).

This concentration on the United Kingdom will also, it is hoped, reflect the experiences and interests of the many thousands of UK construction professionals who want to understand more about the technology available to them and how to adopt and use it most effectively. To help meet this thirst for knowledge, industry writers and academics have produced a steady stream of magazine articles, reports, product surveys, briefings, 'how to' guides and case studies over the past four or five years, but no single publication has attempted to draw all this information together. This book seeks to fill that gap.

Moving on from this chapter's generic definition of the technology, Chapter 2 resolves why collaboration and its supporting technologies have become important now. By the end of 2004, well over 100,000 UK industry professionals had used one or more of the leading systems, and this user community was expected to continue growing for some years. Subsequent chapters are therefore intended to help readers understand, and make informed decisions about, the technology and the relative merits of the various vendors and their services. The benefits (and costs) of using the technology are illustrated by case studies, while pragmatic considerations such as the legal implications, connectivity requirements and training needs are also covered. The book concludes by looking at how collaborative technology might develop in the future.

1.1 What is collaboration?

At the start of the twenty-first century, new internet-based collaboration – or c-commerce – technologies were widely viewed as a major growth sector by IT analysts such as the Butler, IDC, Forrester and Gartner groups and by business process re-engineering gurus such as Michael Hammer (2001) and James Champy (2002).

They recognised that for many corporations, their greatest asset is their 'knowledge capital' ('the value generating asset of a business that includes know-how, ideas, databases and goodwill' – Davis 2003), and the key challenge is to maximise this asset's value while effectively controlling the management of this information. Most knowledge workers tend to work with others to complete tasks, collaborating internally with fellow employees, and/or externally with customers, suppliers, etc. Collaboration takes place at multiple levels, from small peer groups with immediate colleagues to multi-disciplinary project teams, from company-wide activities with members drawn from different grades of the organisation's hierarchy to those that extend beyond the enterprise to become inter-enterprise collaboration. In short, collaboration takes many forms and is required in just about every business process.

To collaborate, says the Oxford English Dictionary, is to 'work jointly (with) esp. at literary or artistic production' (and, perhaps particularly appropriate to the often-adversarial atmosphere of traditional construction projects, it adds: 'to co-operate with the enemy'). This is to look at the term as it applies culturally, but it has in recent years – as we shall see – also gained a technology dimension.

If one looks at the cultural use of the term, some management writers have focused on the creative element. Schrage (1990), for example, defines collaboration as: 'the process of shared creation: two or more individuals with complementary skills inter-acting to create a shared understanding that none had previously possessed or could have come to on their own'.

As this definition suggests, successful collaboration is a process of value creation that cannot be achieved through traditional, often hierarchical structures. However the vital communication takes place – whether face-to-face or virtual – it tends to require the giving and receiving of feedback in an atmosphere of mutual trust and respect between all the interested participants, each specialist in their own fields. This feedback will often result in reassessment of an initial idea, and as the collaborators develop a shared sense of what they are trying to achieve, the outputs can be greater than the sum of all individuals' expertise and knowledge inputs.

In an AEC context, Kalay (1999) defines collaboration as: 'the agreement among specialists to share their abilities in a particular process, to achieve the larger objectives of the project as a whole, as defined by a client, a community, or society at large'.

It is perhaps worth emphasising, too, that true collaboration requires participants to set aside any self-interest or belief that, by their professional background, training, role or project responsibility, they are somehow superior to other members of the team. In the construction context, this was underlined by the UK Strategic Forum for Construction in its *Integration Toolkit* (2003):

> By accepting that there is nothing individuals can do which cannot be done better by a team, collaboration automatically becomes the highest value which can only be reached by truly listening to other people and adding their valuable contribution.
> (Integrated project team (IPT) workbook, section 5.5)

For the purposes of this book, we can absorb these points into a definition of collaboration:

> A creative *process* undertaken by two or more interested individuals, sharing their collective skills, expertise, understanding and knowledge (*information*) in an atmosphere of openness, honesty, trust and mutual respect, to jointly deliver the best solution that meets their common goal.

Of course, this may be somewhat utopian. There will be many instances when what purports to be team collaboration is, in fact, simply a group of individuals going through the motions. They may communicate. They may consult. But the outcome may not actually be the best solution. It may, for example, be the result favoured by a domineering individual. It may be a quick compromise, impacting on quality, but agreed to save time and/or money rather than wholly satisfy the objective. Or its development may have overlooked the interests and inputs of key participants omitted from the 'collaboration' process. We return to this theme when we discuss non-collaborative use of technology later in this chapter.

1.2   Defining collaboration technology

The IT world is awash with technology buzz-words and abbreviations which have been associated with collaboration (enterprise content management, business intelligence, knowledge management, enterprise application integration and business process management are just a few). The term 'collaboration technology' is often used to describe various combinations of software and/or hardware employed to help people collaborate. These include enterprise portals and intranet applications, generic workspace or project team applications, web and video conferencing and online meeting applications, peer-to-peer file-sharing, and real-time instant messaging (IM), to name but a few.[5]

Unfortunately, as one might expect, no standard definition is shared by IT analysts or, more particularly, the various vendors of such collaboration tools. For example, the Butler Group (2003) talks about a 'technology stack' including individual, work-group, enterprise and inter-enterprise collaboration, while the vendors often tend to define the term in a way that suits their product or market – sometimes to the confusion of their customers. For example, some vendors may stress that collaboration requires real-time or concurrent interaction (synchronous), while others will suggest that collaboration is undertaken through a series of interactions with time delays between them (asynchronous).

The difference is fairly straightforward. Synchronous collaboration takes place when participants review and discuss issues in real time. In construction projects, for example, this can take the form of face-to-face conversations, formal and informal site meetings, design reviews, workshops, etc.; if participants cannot be in the same place at the same time, telephone calls, online meetings via webcams and video conferences are all forms of synchronous collaboration. Asynchronous collaboration might involve a design being produced by one person and forwarded to another for review, comment or approval, perhaps by email or fax. After a period of time reflecting on the design (and perhaps some synchronous collaboration – a chat – with colleagues in the office), the recipient may then respond to the originator and suggest improvements resulting, eventually, in a further iteration of that design. On most construction projects, therefore, communications will take place using both forms of collaboration at different times depending on the circumstances – though, with the requirement to think, review and make comments, etc., the onus will tend to be on asynchronous collaboration. But whether synchronous or asynchronous, the key ingredient is the involvement of two or more human beings.

In this respect, arguably, the term 'collaboration technology' may be something of a misnomer. First, if we regard collaboration essentially as a creative process or capability reflecting the roles and responsibilities of the participants concerned, then it is clear that it is *people* that collaborate, not technologies or systems; as the Butler Group (2003) succinctly put it: 'Collaboration is an activity – not a piece of technology'. Using so-called collaboration technology does not necessarily mean one is 'collaborating' (as already suggested, there is more to collaboration than just communicating with fellow project participants). It is, in short, an enabler, a platform that allows collaboration to take place when people are prepared and equipped to do so. Second, albeit a more minor point, 'technology' implies that a single solution is employed. In reality, the technology is actually a combination of several interacting technologies: a computer, a modem, an internet connection, a web-browser and one or more additional software applications.

Individual collaborators need a point of access or contact where they can participate in a process, sharing their collective information with other members of their team. In the above definition of 'collaboration', the words 'process' and 'information' were deliberately emphasised to stress that collaboration involves an *interface* – a shared environment where processes and information can be efficiently and effectively integrated.

This book's initial definition of collaboration can now be expanded to develop a definition for collaboration technology:

> A combination of *technologies* that together create a single shared interface between two or more interested individuals (*people*), enabling them to participate in a creative *process* in which they share their collective skills, expertise, understanding and knowledge (*information*) in an atmosphere of openness, honesty, trust and mutual respect, and thereby jointly deliver the best solution that meets their common goal.

Today's technology market includes numerous different applications that provide a shared virtual workspace that people can use for collaboration, whether the needs are internal or external, enterprise-wide or team-focused, synchronous or asynchronous, etc.[6] The focus of this book, however, is on data-centric applications employed in the AEC market.

1.3 Defining construction collaboration technologies

While 'collaboration' has been a technology buzzword for some years, it has also become an increasingly widely used word within the UK AEC industry since at least the mid-1990s. As we will see in Chapter 2, the term became particularly popular following the 1994 Latham Report, to describe the communication that takes place between team members engaged in partnering approaches to project delivery (see the first part of Chapter 2). Initially, that collaboration was achieved through traditional communication routes (face-to-face, telephone, post, fax, etc.), augmented by email, file transfer protocol (FTP) sites, company intranets, groupware such as Lotus Notes, or (very occasionally) the use of an electronic document management system (EDMS). But wider access to the web during the late 1990s prepared the way – initially in the United States but later in the United Kingdom and other countries – for a new class of web-based software applications that, for want of a better term, are now described as 'construction collaboration technologies'.

Several businesses were launched during the dot.com boom aiming to capture a share of the UK construction collaboration market. The proposition was clear: the AEC industry was (and still is) highly information-dependent, but prone to delay, waste, defects and additional costs due to late, inaccurate, inadequate or inconsistent information. The new web-based applications offered a way to improve the management of this key project information; project times, costs and risks could be reduced; and efficiency, communication and quality improved. However, the technology providers faced the daunting tasks (see the second part of Chapter 2) of not only creating the technologies, but also educating and growing the market – tasks made no easier by the lack of a common terminology.

Perhaps reflecting the different origins and strengths of the various competing suppliers and/or the level of understanding of some of the customers and commentators, a number of different synonyms and abbreviations emerged. These included (in alphabetical order): collaborative extranet, concurrent engineering (CE) environment, construction management system, construction portal, construction project extranet (CPE), construction project network (CPN), data management system, design management system, document exchange, document management system, document pool, drawing management system, enterprise portal, extranet, online collaboration and project management (OCPM) technology, online file storage, project collaboration network (PCN), project collaboration service, project extranet, project hosting, project management platform, project management systems–application service providers (PM–ASP), project portal, project website, virtual project and web-based project management systems (WPMS).

Some of the terms used were inappropriate or potentially confusing. Portal, for example, was often used to describe a website gateway to various sources of information (news, weather, share prices, etc.); the word did not convey the extent to which users could interact with and manipulate the information presented (the same might also be said of website); and an enterprise portal tended to be regarded more as a internal corporate application, not one shared by a multi-disciplinary, multi-company project team. References to document management systems ran the risk that a web-based application would be confused with a (local- or wide-area network) LAN/WAN-based EDMS. Document management systems might also be construed as only managing documents not drawings; similarly, drawing management systems – already employed by some architects – did not tend to be used to manage non-drawings. Construction management, design management or project management systems might be regarded as only relevant to particular disciplines, or, in the latter case, could be confused with project planning and scheduling tools (e.g. Primavera, Microsoft Project, Asta PowerProject, etc.).

As a result, although not strictly correct, the terms 'extranet' or 'project extranet' became among the most widely used.[7] An extranet is frequently defined as a variant of an intranet. The latter is essentially a private network contained within an enterprise that uses internet protocols (IPs) to share company information and computing resources among employees (i.e. an enterprise portal). When sections of this network are made accessible to authorised customers, partners, suppliers or others outside the company, that part becomes known as an extranet – so a section specific to a particular project might therefore be said to be a 'project extranet'. If the collaboration application and its project-relevant data is a sub-division of the customer's intranet, or enterprise portal (or, for that matter, the intranet of the main contractor, principal designer or project manager), then the term would be correct. However (as described in Chapters 3 and 4), in most instances, project-related information is made available from a project-specific website hosted at the software vendor's external facility; it is not part of an intranet. The term 'project extranet', however, does at least convey the key notion that information is private, securely managed and only accessible by authorised team members. A more minor point concerns the word 'project'. Some providers felt this potentially undersold the technology as it might be used to manage multi-project programmes of work or to manage built assets beyond the construction phase when they were no longer regarded as 'projects'.

Semantic debate about what the correct terminology has continued on and off since 2000, but a significant step occurred in 2003. When the founder members of the UK vendors' trade association got together to decide upon a name for their group, they avoided the word 'extranet', opting instead to become the Network of Construction Collaboration Technology Providers – hence this book's use of the term.

The principal features of construction collaboration technology are covered in more detail in Chapter 5, so a short explanation will suffice for now. Broadly, all such systems can be accessed through a computer equipped with a standard computer browser (e.g. Microsoft Internet Explorer, Netscape) and a working internet connection. The same basic functions are common to all. Authorised users, no matter where they are located or when they use the system, can get immediate access to a single, central repository of project data that grows as information about the project or programme (a building, a road, a bridge, a water treatment plant, etc.) is developed by the team. Feasibility studies, budgets, sketches, drawings, approvals, schedules, minutes, photographs, specifications, standards, procedures, virtual reality models, etc., can all be viewed; team members can add comments, issue notices, instructions and requests for information (RFIs), and publish drawings and documents, singly or in batches. Everyone works on the most up-to-date, accurate and relevant information, backed by all the archive material. As shown in Figures 1.1 and 1.2, the single repository offers a more efficient and less complex way to manage communications than through traditional methods.

On a typical construction project, there will usually be dozens of participants, hundreds of drawings and documents, and thousands of information exchange processes. The core project or programme will typically be sub-divided into many smaller packages, phases and/or contracts; similarly, some team members may only be involved in the completion of relatively small tasks; as a result, these individuals may not share the main goal of the project, but their role-specific goals should at least contribute towards the achievement of that main goal. We therefore need to amend our definition to cover some of the principal features of the technology and to allow for these multiple goals.

'Construction collaboration technology' is:

> A combination of *technologies* that together create a single shared interface between multiple interested individuals (*people*), enabling them to participate in creative *processes* in which they can openly share their collective skills, expertise, understanding and knowledge (*information*), and thereby jointly deliver the best solution that meets their common goal(s), while simultaneously creating an auditable electronic record of the people, processes and information employed in the delivery of the solution(s).

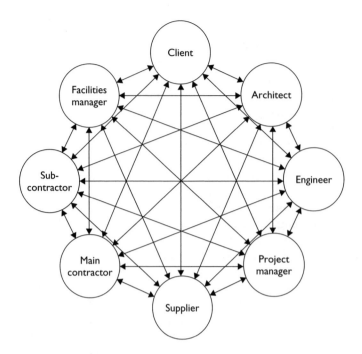

Figure 1.1 Traditional project team communications.

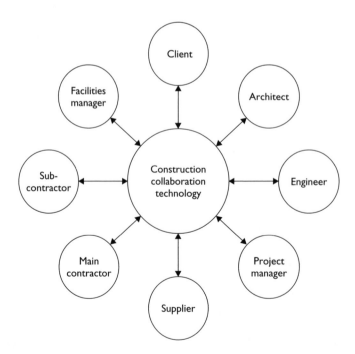

Figure 1.2 Project team communications using construction collaboration technologies.

1.4   Using the technology non-collaboratively

The above definition comes with a warning, though. It deliberately echoes points made repeatedly by vendors, commentators, conference speakers and users since 2000, to the effect that:

1   collaboration requires a combination of people, processes and technology/ information;
2   successful collaboration is 80 per cent people and processes and 20 per cent technology/information (the exact numbers vary – some say it is more like 90/10, for instance – but the stress is always on the people and process issues).

In other words, successful collaboration is much more dependent on the *culture* of the team than it is on the *technology* it employs. Stressing how complex and challenging collaboration can be, Kalay (1999) pointed out how collaboration in the construction sector is different from collaboration in other fields:

> First, it involves individuals representing often fundamentally different professions who hold different goals, objectives, and even beliefs. Unlike collaborators in medicine or jurisprudence, who share a common educational basis, architects, structural engineers, electrical engineers, client, property managers, and others who comprise a design team, rarely share a common educational foundation, and often have very different views of what is important and what is not. Second it involves…'temporary multi-organizations': a team of independent organizations who join forces to accomplish a specific, relatively short-term project. While they work together to achieve the common goals of the project, each organization also has its own, long-term goals, which might be in conflict with some of the goals of the particular project, thereby introducing issues that are extraneous to the domain of collaboration. … Third, collaboration in the building industry tends to stretch out over a prolonged time, even when the original participants are no longer involved, but their decisions and actions still impact the project.
>
> (p. 4)

While the various vendors may like to describe their systems as collaboration technologies, some of them may not, in fact, enable collaboration at all, particularly if the project team is unable or is not prepared to collaborate. For example, a team may simply use the technology as a convenient way to store and distribute information more quickly and cheaply while retaining existing, often opaque project processes, thus maintaining the traditional adversarial construction culture of mistrust and self-interest. There may still be benefits in terms of faster communication, easier access to a central repository, reduced paperwork costs, etc., but more substantial savings of the kind envisaged by Latham, Egan *et al.*, and promised by the technology providers (albeit with caveats), will only come from rethinking and changing the culture of the team and the processes it employs (see further discussion of the technology benefits in Chapter 9).

If a project team persists with old-style contractual attitudes and approaches to project delivery (and some industry estimates suggest that only 30 per cent of the

UK construction industry is actually practising some form of partnering or collaborative working),[8] then it could be resistant to using such IT at all, it may want the technology to replicate its traditional approaches and processes, or it may use the technology for some processes while retaining alternative systems (e.g. separate email transactions, faxes, paper-based correspondence) for others. Where the technology is only partially deployed or is deployed to support processes that are not collaborative in the sense defined so far in this chapter, then perhaps we need to offer an alternative definition. The word collaboration would clearly no longer be applicable, so perhaps *communication* is more appropriate:

'Construction *communication* technology' is:

> A combination of *technologies* that together create a shared interface enabling certain authorised individuals (*people*) to participate in certain *processes* through which they fulfil contractual obligations by contributing their skills, expertise, understanding and knowledge (*information*), while creating an electronic record of the people, processes and information employed in that element of the delivery of the solution(s).

This revised definition recognises that there is no longer a single interface between interested parties; that some parties may be excluded from certain processes; that some will be simply be providing information they are contracted to deliver; that there is little or no collaborative culture; and – perhaps most fundamentally – that there is no longer a single complete and auditable record of the project's delivery.

1.5   Chapter summary and plan of Chapters 2–10

1   This chapter was intended to set the scene for the remainder of the book by answering the question: 'What is collaboration?' It has briefly described the rapid growth of internet-based IT during the past two decades before narrowing its focus to the use of collaborative applications specific to the AEC industry in the United Kingdom. C-commerce has been an IT industry buzzword for some years and 'collaboration' has become a widely embraced, if sometimes abused, term within both the IT and construction industries. Chapter 2 develops this theme further.

2   *Why collaborate now?* Chapter 2 looks at the recent history of partnering and collaborative working, and relates this to the initially parallel, but then increasingly convergent development of technologies to support such collaborative approaches.

While some clients and/or project team members may have previous experience of using construction collaboration technologies, in many cases a team may need to select from what, at first sight, appears a confusing array of different companies and software systems. The bulk of this book is therefore devoted to expanding on the key issues relating to selection. Construct IT's guide to managing e-Project information (2003, p. 15) suggested the following criteria as a basis for pre-selection in order of priority:

- functionality offered and available without modification;
- services offered and service availability including specialist training, user and application support;

- performance issues given particular operating environments;
- track record in e-Projects including reference sites;
- financial stability of the software company;
- price.

This book covers all of these criteria, although not necessarily in this sequence. Arguably, the consequences of a provider's financial failure may be quite severe, so the background of the software vendors is considered in some detail before we look at system features and other factors.

3   *Who provides collaboration systems?* Chapter 3 is devoted to developing an understanding of the key generic differences between the various vendors of construction collaboration technologies.

4   *Who hosts the systems?* Chapter 4 aims to help the reader learn more about hosting and the pros and cons of the application service provider (ASP) model.

5   *What do collaboration systems do?* Chapter 5 focuses on software functionality, and the particular challenge of managing and enabling access to graphical data, plus service support issues.

6   *How do we connect to the systems?* Chapter 6 relates to how end-users connect to and achieve satisfactory speed performance from typical collaboration systems.

7   *What are the legal issues?* Chapter 7 explains how successful implementation will require careful attention to the various legal issues.

8   *What are the vital human factors?* Chapter 8 looks at the people and process issues, including price, critical to successful selection and introduction of the technologies into project team environments.

9   *What are the benefits of using the systems?* Chapter 9 looks at perhaps the single most important positive influence on use of collaboration systems: the benefits they can deliver.

Finally, having discussed in detail the wide range of selection criteria and related variables, the book concludes with a look to the future:

10 *What does the future hold for collaboration systems?* Reflecting the continued evolution of the AEC industry, its IT tools and the telecommunications at its disposal, Chapter 10 speculates about potential developments that may influence the future use of collaboration systems.

Chapter 2

The convergence of culture and technology

This chapter:

- describes the emergence of more collaborative attitudes and styles of working within UK construction project teams;
- shows how technologies developed to increase individual productivity have gradually become capable of supporting collaborative working and information sharing;
- relates the above two trends, and discusses factors affecting the rate of technology adoption.

The increased profile of online collaborative working in the United Kingdom during the early years of the twenty-first century is largely due to the convergence of two long-term trends. The first relates to the culture and working practices of the UK AEC industry; the second relates to the availability and use of information and communication technologies (ICT) and the resulting growth in mobile and remote working.

This chapter looks at the history of these two trends, and then briefly discusses some of the key issues that affected – and in some cases continue to affect – the rate of adoption of the technology.

2.1 From partnering to collaborative working

For many current construction industry professionals, the buzzword 'partnering' dates back to 1994 and the publication of Sir Michael Latham's report *Constructing the Team*. However, Latham simply marked the latest in a long series of industry reports (e.g. the Simon Report in 1944 and Banwell 20 years later) urging contract reform – a series that has since been extended, as we shall see. His recommendations also paralleled similar initiatives in other industry sectors. The automotive industry, for instance, had already been paying close attention to 'lean thinking' in manufacturing: *The Machine that Changed the World* (Womack *et al.* 1990) documented how Japanese car maker Toyota had revolutionised both its supply chains and its distribution networks during the 1970s and 1980s. The UK offshore oil and gas exploration industry had also begun to move towards less adversarial working methods through initiatives such as Cost Reduction Initiative for the New Era (CRINE).

As Crane and Ward (2003) discuss, there are construction industry practitioners that suggest there is nothing new in partnering (it's how we have always done things in our firm). Indeed, Alan Crane, chairman of Rethinking Construction, tells audiences that partnering approaches arguably predated Latham by some 60 years: retailer Marks & Spencer worked closely with contractor Bovis on successive projects from the 1920s to the 1980s, while the collective need to 'dig for victory' during the Second World War saw a temporary suspension in many adversarial relationships as contractors looked to make vital contributions to the government's war efforts. Some client organisations, notably in local government before privatisation and outsourcing initiatives took hold, also had close-knit in-house teams with designers and project managers drawn from their own architectural/engineering departments and work undertaken by their own direct labour teams.

However, in 1994 there was no widely shared understanding of what exactly was meant by partnering, and the existing, somewhat woolly thinking did not prescribe what actions needed to be taken to move forward. Since then, though, the UK construction industry has begun to develop and disseminate increasingly precise definitions of the term and to identify some of the activities and – perhaps more importantly – the attitudes necessary to implement a more collaborative approach to construction procurement.

2.1.1 The Latham era

Latham made 53 recommendations to change industry practices, to increase efficiency and to replace the bureaucratic, wasteful, adversarial atmosphere prevalent in most construction projects with one characterised by openness, co-operation, trust, honesty, commitment and mutual understanding among team members:

> Partnering includes the concepts of teamwork between supplier and client, and of total continuous improvement. It requires openness between the parties, ready acceptance of new ideas, trust and perceived mutual benefit.... We are confident that partnering can bring significant benefits by improving quality and timeliness of completion whilst reducing costs.
>
> (para 6.45, p. 62)

> Partnering arrangements are also beneficial between firms.... Such arrangements should have the principal objective of improving performance and reducing costs for clients. They should not become 'cosy'. The construction process exists to satisfy the client. Good relationships based on mutual trust benefit clients.
>
> (para 6.46, p. 62)

A year later, he wrote the foreword to a best practice guide to partnering in construction from the Reading Construction Forum (RCF) (*Trusting the Team*, Bennett and Jayes 1995). This defined partnering as:

> a management approach used by two or more organisations to achieve specific business objectives by maximising the effectiveness of each participant's resources. The approach is based on mutual objectives, an agreed method of problem resolution and an active search for continuous measurable improvements.
>
> (p. 2)

The RCF also emphasised that it was not all about cost-cutting: 'partnering can also improve service quality, deliver better designs, make construction safer, meet earlier completion deadlines and provide everyone involved with bigger profits' (p. iii).

Within a notoriously conservative industry, some contractors and consultants enthusiastically embraced the concept, recognising the value of establishing long-term relationships with customers and other members of the supply chain (why, after all, should the construction industry be any different from other sectors, where relationship marketing was already a familiar concept?). The RCF and others strongly advocated moving beyond 'project partnering' to 'strategic partnering' or 'alliancing' to cover just such an eventuality. Such relationships had to be preferable, they argued, to the continued reliance on complex business relationships that were often relatively short-lived. Why assemble a large, expensive, geographically dispersed, virtual organisation including the client, architects, engineers, contractors, suppliers, construction managers and other professionals only to disband it once the project is handed over? While project partnering might deliver cost savings of between 2 per cent and 10 per cent, the RCF suggested that it was only through more strategic, long-term arrangements that substantial cost savings (of up to 30 per cent) could be achieved.

Clients too could see the advantages. Innovative organisations such as airports operator BAA and retailer Sainsbury's justified building more long-lasting, strategic relationships on the grounds that they were also capturing information, experience and best practice. They realised that information created during project delivery was a valuable 'whole life' asset that could be used to enable better planning, continuous performance improvement and risk reduction across their current and future property portfolios. After a long and detailed review process, BAA, for example, established long-term framework agreements with various groups of suppliers to deliver key parts of its infrastructure.

Many other businesses – particularly subcontractors, suppliers and manufacturers – remained sceptical, but the industry did start to move forward. The pan-industry Construction Industry Board (CIB) was established to drive forward Latham's change agenda, and under Don Ward's leadership in 1997 it published a report, *Partnering the Team*, which defined partnering in a way that stressed the need for strong, conscious management of the process: 'Partnering is a structured management approach to facilitate teamworking across contractual boundaries. Its fundamental components are formalised mutual objectives, agreed problem resolution methods, and an active search for continuous measurable improvements'.

Crane and Ward (2003) emphasise the terms 'structured', 'formalised', 'agreed', 'active' and 'measurable' to differentiate genuine partnering from other forms of cooperation – as the CIB did: '[Partnering] should not be confused with other good project management practice, or with long-standing relationships, negotiated contracts or preferred supplier arrangements, all of which lack the structure and objective measures that must support a partnering relationship'.

Alongside the CIB, Construct IT was established to coordinate and promote innovation and research in IT in UK construction and to act as a catalyst for academic and industrial collaboration. The Construction Best Practice Programme (CBPP, and its sister programme, IT Construction Best Practice, ITCBP) was created to provide guidance and advice, enabling UK construction and client organisations to gain the knowledge and skills required to implement change. And the Design Build Foundation (DBF) was launched in 1997 as a catalyst for change, drawing together forward-thinking construction industry customers, designers, contractors, consultants, specialists and manufacturers, representing the whole construction supply chain.

Latham and the ensuing industry initiatives had excited a lot of interest but uptake remained slow. The next step was backed by government action. In October 1997, the Deputy Prime Minister John Prescott commissioned the Construction Task Force, chaired by a former chief executive of Jaguar Cars, Sir John Egan, to adopt the clients' perspective (advocated, as we have seen, by Latham) and advise on opportunities to improve the efficiency and quality of the UK construction industry's service and products, to reinforce the impetus for change, and to make the industry more responsive to the needs of its customers.

2.1.2 The Egan era

Informed by experiences in other industries (notably manufacturing), the Task Force report *Rethinking Construction* (1998) – the Egan Report – endorsed much of the progressive thinking already under way. *Rethinking Construction* acknowledged that its foundation was the Latham Report, which the CIB had helped implement by promoting a focus on client value, partnering and standardisation. It sought to achieve a further step change in performance through eliminating waste or non-value-adding activities from the construction process, and identified five key drivers of change:

- committed leadership;
- a focus on the customer;
- integrated processes and teams;
- a quality driven agenda;
- commitment to people.

Having put the client's needs at the very heart of the process, it advocated an integrated project process based around four key elements:

- product development;
- project implementation;
- partnering the supply chain;
- production of components.

Existing industry bodies such as the CIB, CBPP and the DBF enthusiastically incorporated the Egan agenda into their activities, and were augmented by a new industry organisation, the Movement for Innovation (M4I). This was established, with Alan Crane as the first chairman, to stimulate the application of the ideas of *Rethinking Construction*, most notably through 'demonstration projects', but also through regional 'cluster groups' or best practice clubs.

Other organisations were also formed to promote the new agenda. The short-lived Confederation of Construction Clients was established to support the industry's clients; its 2000 charter urged commitment to four key areas:

- client leadership (including the adoption of partnering approaches);
- integrated teams;
- a quality agenda;
- respect for people.

Focusing change efforts in particular sectors of the industry, the Housing Forum, the Local Government Task Force and the Government Construction Clients Panel were also established. With public sector work accounting for some 40 per cent of the construction industry's output, local and central government were seen as key areas, and even before *Rethinking Construction* was published, some government departments were beginning partnering experiments.

The 'Building Down Barriers' initiative was established in January 1997, funded by the then Department of the Environment, Transport and the Regions and by Defence Estates within the Ministry of Defence. This had three overall objectives:

1   To develop a new approach to construction procurement, called Prime Contracting, based on supply chain integration.
2   To demonstrate the benefits of the new approach, in terms of improved value for the client and profitability for the supply chain, through running two Pilot Projects....
3   To assess the relevance of the new approach to the wider UK construction industry.

The pilot projects were new training and sports centres at the army garrisons of Aldershot and Wattisham. Prime Contractors Amec and Laing, respectively, led supply chains of designers, specialist contractors, materials suppliers and component manufacturers, and worked collaboratively with the client to develop a design, then deliver the facility and manage it for a further period of time. The initiative yielded a new procurement process and a 'tool-kit' to support it, largely captured in the *Handbook of Supply Chain Management* (Holti *et al.* 2000).

In March 1999, the UK government's *Achieving Excellence in Construction* initiative was launched to improve the performance of central government departments, their executive agencies and non-departmental public bodies as clients of the construction industry. The initiative set out a route map with performance targets under four headings:

- management;
- measurement;
- standardisation (within which 'IT and standardised document handling' was explicitly identified as a key area);
- integration.

Targets included the use of partnering and development of long-term relationships (see OGC 2003a). Against this background, other government departments began to recognise the impact partnering could make and to promote the approach (e.g. CABE/Treasury 2000; National Audit Office 2001).

Positioning itself as the champion of collaborative working, in 2002 the DBF amended its mission to reflect the new focus: 'to develop and promote integrated design and construction through collaborative working to deliver customer satisfaction'. With the Warwick Manufacturing Group, the DBF (by now about to merge with RCF to form a new supply chain body, Collaborating for the Built Environment (BE)) also established the Collaborative Working Centre (CWC Ltd) to develop and promote the principles of supply chain integration and collaborative working.

BE was launched in October 2002, with Don Ward as chief executive. By combining and developing the work of the DBF and RCF, it was intended to create a single, industry-owned body devoted to radical improvement. Its membership includes many of the top-spending clients in the United Kingdom and a balance of designers, consultants, contractors, specialists and suppliers (including, in due course, some of the collaboration technology providers), all committed to sharing and putting into practice new ideas. In this respect, BE was a major force for change within the UK construction industry.[1]

2.1.3 Accelerating change, and beyond

Almost simultaneously, the Strategic Forum for Construction – successor to both the CIB and the earlier Task Force, but also chaired by Sir John Egan – produced a follow-up to *Rethinking Construction*. The new report, *Accelerating Change* (Egan 2002), set rigorous targets for the construction industry, challenging it to provide maximum value for clients and end users. It echoed the calls for greater integration, setting a key target: by the end of 2004, 20 per cent of construction projects by value should be undertaken by integrated teams and supply chains, rising to 50 per cent by 2007. It also committed the Strategic Forum to producing its own 'toolkit' by April 2003 to 'help clients, and individual supply side members, assemble integrated teams, mobilise their value streams and promote effective team working skills and then produce an action plan to promote its use'.

Perhaps for the first time, a government-backed report on the construction industry specifically highlighted the key integration catalyst role of IT. *Accelerating Change*'s vision saw a construction industry characterised by (among other things): 'Integrated teams, created at the optimal time in the process and *using an integrated IT approach*, that fully release the contribution each can make and equitably share risk and reward in a non-adversarial way' (p. 10, emphasis added).

Client leadership, integrated teams and tackling 'people issues' were seen as powerful drivers for change, but *Accelerating Change* acknowledged that there were several other cross-cutting issues that could act as enablers or barriers to change. One was 'IT and the internet':

> IT and E-business, as enablers, have already radically transformed many operations in the construction sector and there is still a vast potential for more. IT can deliver significant benefits for designers, constructors and building operators. Deriving the maximum benefit from introducing IT solutions will not, however, be easy. There is the potential to drastically reduce infrastructure cost behind the tendering side of the industry by adopting the wider use of the Internet and e-procurement specifically.
>
> The widespread adoption of e-business and virtual prototyping requires the construction industry to transform its traditional methods of working and its

business relationships. Key barriers to this transformation include organisational and cultural inertia, scale, awareness of the potential and knowledge of the benefits, skills, perceptions of cost and risk, legal issues and standards....

(paras 7.8 and 7.9, p. 36)

(We will return to some of these barrier issues elsewhere in this book when we look at factors influencing the selection and use of collaboration technology.)

The need for more integrated teams was already a recurring theme in industry literature. Clive Cain (2001), a prominent advocate of industry reform, urged teams to aspire to the creation of the 'virtual firm': a totally integrated design and construction supply chain (see also Crane and Saxon 2003, p. 57). Echoing *Accelerating Change*'s aforementioned vision of 'integrated teams, ...using an integrated IT approach' (Egan 2002, p. 10), one characteristic of such a 'firm', Cain (2002) told a London conference, would be their use of a system for 'instant exchange and update of data and drawings...[allowing] design input from all supply-side firms'.

Two toolkits were launched during 2003 and both also made explicit reference to the importance of IT to support partnering. First, the Office of Government Commerce launched its Successful Delivery Toolkit in July and pushed the *Achieving Excellence* initiative forward with a series of procurement guides. A total of 11 procurement guides were eventually produced including booklets on project organisation, procurement lifecycle, risk and value management, procurement and contract strategies, whole life costing, health and safety and sustainability. The very first paragraph of Procurement Guide 05 (*The Integrated Project Team: Teamworking and Partnering*) underlined the importance of partnering:

> client and suppliers working together as a team can enhance whole-life value while reducing total cost, improve quality, innovate and deliver a project far more effectively than in a traditional fragmented relationship that is often adversarial. Collaborative working should be a core requirement for each element of every project. Putting it into practice through teamworking and partnering requires real commitment from all parties involved, but brings benefits that far outweigh the effort involved.
>
> (OGC 2003b, p. 2)

The guide (p. 5, italics added) listed six key principles of partnering:

- early involvement of key members of the project team;
- selection by value, not lowest price;
- *common processes such as shared IT*;
- a commitment to measurement of performance as the basis for continuous improvement;
- long-term relationships in the supply chains;
- modern commercial arrangements based on target cost or target price with shared pain/gain incentivisation.

The second toolkit (referenced briefly in the above OGC guide, and helpfully using the same IPT terminology) was the Strategic Forum for Construction's Integration

Toolkit, published in late 2003. In line with *Accelerating Change*'s overt acknowledgement of the growing importance of IT and with the *Achieving Excellence* guide, IPT Workbook 5 included a specific section devoted to communication, including the use of 'Integrated systems'. It said: 'Modern technology should be the catalyst for common systems and open information channels for use throughout the IPT for: design and drawings, project planning and resourcing, safety management, value management, cost planning, cross-disciplinary training.' The suggested tools and techniques included communication platforms and protocols, with Building Information Warehouse (BIW), BuildOnline and Asite among the listed providers of integrated systems.

As well as toolkits, there were also action plans. For example, BE's November 2004 annual conference focused on research by Business Vantage (2004a), which developed a 10-point action plan that customer and supplier organisations were urged to follow to improve the delivery of public sector projects.[2] This underlined many of the points raised in the OGC procurement guides. For example, its final three actions related to:

- *Supply chains* Leaders of supplier organisations were urged to 'commit to investing in the development of your key suppliers', while customers were urged to 'keep successful teams together'.
- *Early involvement* Suppliers should 'communicate early' while customers should keep suppliers in the loop about possible needs.
- *Continuity of people* Suppliers were told: 'In a people business don't let your main asset walk away', while customers were cautioned: 'The team has just gone up a steep learning curve...why start again? The same logic applies to your selection of suppliers' (Business Vantage 2004a, pp. 6–7).

2.2   The growing use of IT

By 2003, then, the future direction of the UK construction industry was being tied very much to the adoption and implementation of more collaborative and integrated methods of working. The key role of IT was also becoming increasingly apparent, and was being repeatedly stressed by industry change initiatives. But it would be wrong to view the emerging importance of IT as a recent development. Since at least the 1960s, like most other sectors of British industry and commerce, the construction industry has been keen to use new IT tools where appropriate (see Sun and Howard 2004). It may have lagged behind some other sectors in the speed of its adoption of some technologies,[3] but it would be wrong to describe the industry as entirely techno-phobic.

2.2.1   A preference for paper

In a very price-sensitive industry, as the cost of hardware and software has reduced, contractors, consultants and suppliers have been enthusiastic to adopt new technology that would allow them to cut costs and waste, increase speed and productivity.

However, it is sobering to realise just how quickly the industry has moved forward and how much the pace of change has accelerated over the past few decades. To Victorian greats such as Brunel or Pugin, the working practices of the professional civil engineer or architect of the late 1970s (with the exception of the telephone, calculator or telex) would have been very familiar. Visitors to the offices of a multinational engineering consultancy, for example, would have seen many of its engineers and technicians hard at work on traditional drawing boards; its reprographics room or post-room would have been busy with staff folding or rolling copies of drawings ready for delivery by post or courier; and numerous support staff would have been hard at work generating reports, correspondence and invoices on mechanical typewriters.

Ten years later, despite high software and hardware costs,[4] often temperamental hardware, steep learning curves for their operators, and major changes to working practices and processes, computer-aided drafting/design (CAD) workstations and plotters had begun to replace the designers' drawing boards, and, yet, for some years their outputs were still largely delivered as physical drawings (it is worth noting, along the way, that it took CAD a decade or so to become accepted by most as the standard way of working within the AEC industry). Fax machines (low cost, no learning curve and no process changes) had become increasingly important – used, for example, to speed up the transmission of design queries, perhaps with a paper copy to follow in the post. Word-processors (initially expensive, but small learning curve and, again, little process change) had begun to replace the latest electric type-writers, and the first, and by modern standards, cumbersome mobile telephones were beginning to change conceptions of the mobile workforce.[5]

Nonetheless, key aspects of the inefficiency that provoked Latham's report (1994) included, first, the industry's reliance on slow, and often labour-intensive paper-based processes to share project information, second, the lack of integration between the design and construction processes, and, third, the short-lived nature and insularity of most commercial relationships.

By the mid to late 1990s, while many construction professionals had computers on their desks, paper documents, drawings and correspondence (with all their attendant amendments, many out-of-date before they reached recipients by conventional means of delivery), remained the main communication media. During even very modest projects, project teams created, copied, distributed and stored huge volumes of information. New electronic capabilities – including CAD, spreadsheets, word-processing, desk-top publishing and database applications – might have facilitated the production of information, but the key industry tendency was almost always to turn the end product back into paper (e.g. while 79 per cent of product specifications were electronically produced, 91 per cent were still distributed on paper – Building Centre Trust 1999). Even if postal services and couriers were being supplemented electronically by faxes and then email, team communications on the vast majority of AEC projects was still achieved through traditional means: face-to-face meetings, telephone voice calls (land-line and mobile) and paper-based communications.

This is perhaps no surprise: paper has a clear and familiar 'look and feel' that allows ideas to be conveyed and read quickly, it is cheap to produce and distribute, and can easily be archived without worrying about whether it will be readable following a software upgrade. Despite the apparent advantages of going 'digital', there remained, as Duyshart (1997) pointed out, no hurry to start working in a different medium:

> Today, digital documents are generally regarded as having far greater functionality and usability features than their paper predecessors. Digital documents are capable of being easily reused for different purposes, searched by title and content,

managed in customised work environments, delivered over vast distances almost instantaneously, and accumulated in valuable information repositories. Yet despite these advantages, few users seem accustomed to the capabilities and use of this relatively new medium.

(p. 3)

This preference for paper also reflects an industry that, in places, can lag behind others in embracing new IT or adopting the latest management thinking. Even as late as 2001, there were still organisations that could justify not using new technology: the Construction Confederation (2001) found small contractors commenting: 'Although we have email and Internet facilities, we find most of the people we want to deal with do not, or prefer phone, fax or mail', and a medium-sized contractor insisting: 'We do not want email in the office because it will waste a lot of staff time'. Similarly, the DTI benchmarking study (2004) found construction businesses had one of the lowest levels of connectivity and networking technologies.

The traditional sequential approach to construction procurement was also inefficient. For most projects, a design team would spend a considerable time developing its proposals to an advanced level of detail before involving a constructor and making a start on site. Even though 'lean thinking' was making in impact in other industry sectors such as manufacturing, construction project teams rarely, if ever, shared information with the rest of their supply chains; specialist suppliers, contractors and sub-contractors were, thus, unable to contribute to design development from the outset.

The short duration and insularity of many project team relationships was also a factor in the industry's inefficiency. A team was usually assembled only for the duration of an individual project, and often professionals would not be co-located until the project started on site (even then, the site team would normally represent only a small portion of the total team involved). Whether office- or site-based, all team members would be constrained by the available technology from sharing information freely, and the status of the project would vary according to the perspective of the individual concerned and the information he or she had received. And at the end of the project, of course, the team was disbanded, dispersing the collectively acquired knowledge.

Nonetheless, Latham's report was timely insofar as it stimulated much discussion about how teams should collaborate at a time when the technology to enable that collaboration was just beginning to emerge.

2.2.2   The UK construction and the WWW

When the Latham Report was published in 1994, for example, the WWW – developed initially as a collaborative platform for US government purposes – was still in its infancy. Few businesses had internet connections,[6] few (if any) UK AEC businesses had websites (indeed, the WWW then amounted to just 3,000 websites) and research into other ways to use 'the information superhighway' was just beginning. That year, the Department of the Environment funded a research project, undertaken by the Building Research Establishment, Newcastle University and Nottingham software developer Engineering Technology, to develop the industry's first information portal website: the BIW.[7] This was launched at the beginning of 1995 to a sceptical industry and showcased on the DoE stand at the Interbuild exhibition in Birmingham in November 1995. To put this in context, the WWW by then had grown to just 25,000 websites, and there were just 16 million internet users worldwide.

If the potential value of the internet to the construction industry was not immediately clear to industry professionals, it was already being recognised by IT gurus such as Microsoft's Bill Gates (1996):

> The Internet has a huge potential as it relates to construction. This is an industry that continually moves information back and forth between offices and remote job sites. Pulling together even a simple straightforward project now requires the interaction of hundreds of people and thousands of documents. Today's communication challenges are incredible.

However, by the time *Rethinking Construction* was published (Egan 1998), construction websites – often little more than online 'brochureware' – were becoming commonplace (there were now 4.2 million websites worldwide).[8] In some sectors AEC businesses with internet connections now exceeded those without (e.g. Fedeski and Sidawi (2002) found more than 60 per cent of architects practices had internet connections by the end of 1998). And email was fast superseding the fax as the preferred medium for rapid written communication.

In the late 1990s, the personal computer (PC) and Microsoft's Windows operating system had heralded a new era in personal productivity. Almost all AEC professionals had become computer users, particularly for word-processing, spreadsheet work and – in the design community, at least – CAD, with high rates of email use and growing levels of web access (by 2004, there were 830 million internet users worldwide[9]). But, for many users, the technology was still not yet used to its full potential and capacity. The client–server basis of most IT systems, and their lack of integration, did not promote collaboration with external partners; telecommunications links were often expensive, slow and/or overloaded; and computer processor speeds were barely, if at all, adequate. Beyond technical considerations, training requirements, lack of investment and a preference for paper were users' key concerns (Building Centre Trust 1999), along with continued industry reservations about the legal admissibility of such electronic communications.

This gradual, practical adoption of new technology reflects a conservative, risk-averse industry in which projects frequently take months, even years, to move from inception to completion. Quite justifiably, IT advances would not usually be introduced part way through a project; conventional client–server applications tended to be expensive to deploy at temporary, geographically dispersed project sites; and, even if a project was still at inception stage, clients and team members would normally want reassurance that the new technology was already tried, tested and proven elsewhere, ideally on a project of a similar type and scale.

This makes it all the more curious then that, briefly, some of the best-known names in UK construction made what some would consider ill-judged efforts to capitalise upon the e-commerce revolution.

2.2.3 The UK construction goes dot.com dotty

Around the world during the dot.com boom of the late 1990s, hundreds of businesses were started in an attempt to capitalise on new e-business opportunities. Being first to market and grabbing as big a slice of the market as possible were seen as essential, and business-to-business (B2B) e-marketplaces or trading exchanges were among the most common ventures to be launched. Some were speculative internet 'pure-plays' (construction-focused ones included constructionhub.co.uk and jackhamma.com); others were established industry players looking to find new routes to market.

Many existing UK construction businesses began to investigate use of the internet as a means of transacting business. In the opening months of the twenty-first century, for instance, a succession of major companies announced their involvement in new e-procurement ventures. In April 2000, Aggregate Industries, Alfred McAlpine, BPB, Pilkington and RMC came together to back Mercadium: 'a neutral electronic market place for building materials that will be open to all industry participants'. In June 2000, leading international engineering and construction companies AMEC, Bovis Lend Lease, Hochtief, Turner and Skanska announced they were launching AECVenture, 'an independent marketplace open to the global AEC industry'. A month later, in July, five British companies – the aforementioned AMEC and Bovis Lend Lease, plus Balfour Beatty, Kvaerner and John Laing – announced they were joining with AECVenture to create Arrideo, 'a business-to-business (B2B) exchange serving the entire UK construction and engineering industry'. Other would-be Pan-European ventures mushroomed (e.g. EU-supply was founded in Scandinavia in 1999 and B2Build was launched in Brussels) while other businesses attempted to focus on particular sectors of the industry (e.g. Build Europe, trading in the United Kingdom as thebuilding-site.com, targeted the European house-building market).

However, by the end of 2001, the dot.com bubble had well and truly burst and the heady optimism of the late 1990s had evaporated – as had almost all of these B2B e-marketplace ventures. It is not hard to work out why they failed. On the supply side, sellers were concerned that customers might team up to aggregate their spending power and drive down prices. Buyers, on the other hand, were worried that commercially sensitive information about their purchasing habits might be revealed to competitors, and – particularly in the post-Latham/Egan, pro-partnering era – many were reluctant to erode the close relationships that they had developed with key suppliers. And the backers of the various ventures almost certainly underestimated the costs and technical challenges involved in concentrating huge volumes of data into central e-procurement exchanges.

2.2.4 Towards collaboration technology

While the construction e-marketplaces disappeared, other construction e-businesses were finding their feet with a more modest proposition: collaboration.

The rationale was clear. With the AEC industry being so information-dependent, sharing accurate, timely information was critical for all participants. Wasted time and cost can almost always be traced back to poor co-ordination caused by late, inaccurate, inadequate or inconsistent information – sometimes a combination of all four.

Moreover, most of the industry's IT applications did little to improve matters, being designed as stand-alone tools (e.g. CAD was separate to programme management, which was separate to cost control) with little integration between them. With team members increasingly widely dispersed and also increasingly mobile, what was needed was some means to communicate, centralise and share that information more quickly and efficiently.

Of course, email offered one possibility. Sometimes described as the internet's first 'killer app',[10] email enabled many construction professionals to send drawings, documents and other files as attachments almost instantaneously (one CAD managers survey (Davies 2004) found email was the most popular way to issue digital information, used in 76 per cent of cases). In the absence of other tools and technologies, email also became a vehicle for different types of interactions (e.g. threaded discussions, chat, scheduling, etc.). But email use can quickly turn into abuse as participants are added to distribution lists indiscriminately, overloading inboxes, requiring disproportionate management, and providing no easy way to centrally store, retrieve, manage or track the interactions between project team members – and that assumes that the email system is reliable.[11] As email use became increasingly universal in the early twenty-first century, construction professionals were just as likely to be affected by problems such as viruses and 'spam'. By mid-2004, it was estimated that as much as 70 per cent of email was spam (with the problem further compounded by virus alerts automatically generated by anti-virus programs).[12] These problems, of course, will also have implications for any collaborative processes that rely on email, and have contributed to the growing use of alternative communication routes such as IM.

Project communications could also be sent using FTP, one of the earliest internet applications (now over 30 years old) which allowed users to share online repositories of files. Using an FTP application, authorised users could send files via the internet to a specified remote server which could then be accessed by other authorised users and the files downloaded. At a time when many users were still accessing the internet via slow dial-up modems, FTP offered an alternative to sending large files as email attachments. A file could be uploaded once to one location, then downloaded at a time convenient to the end-user(s). However, most FTP applications were not easy to use (and were therefore prone to user error) and offered little, if any, workflow management or auditability.

Groupware such as Lotus Notes allowed teams to share documentation easily, but the sheer proliferation of Lotus Notes databases, and the need to allow wider access to them, soon presented problems.

Some construction businesses were also beginning to experiment with intranets: company-specific private websites that could be used to store key information, usually corporate but sometimes project-related. However, these sites tended to be static and most businesses were reluctant to open them up, even in narrowly restricted areas, to project partners from external businesses.

During the 1990s, EDMSs had become increasingly common in manufacturing. Users normally accessed these server-based systems via LANs or WANs, and were able to check in/check out documents so that an audit trail of who has done what to them and when can be maintained, but the applications themselves could be expensive to purchase, implement and support, and only managed information 'inside the firewall'. As such, EDMSs were largely unsuitable for AEC projects where project margins were already thin, where teams tended to be temporary, geographically dispersed and multi-company, and where many users – especially at smaller sites – might not have network connections and only low bandwidth internet connections.

Similar arguments could also be applied to projects where a dedicated WAN might be set up to connect all consultants, contractors and suppliers to a shared drive on a central server hosting all the project documents. In both cases, the investment might perhaps only be justified on particularly large and/or high value projects of long duration. Moreover, a network-based system – unless some kind of information management application was deployed – might suffer from poor folder structures, limited flexibility, the risk of files being overwritten, deleted or mis-filed, poor security, etc.

However, the development of a new breed of business applications – collaboration, or c-commerce, technologies – during the mid-1990s heralded some major changes in the construction industry. Applications were initially developed on a client–server architecture but alternative, internet-based technologies soon followed. These were spearheaded initially in America by firms such as MPInteractive (now e-Builder), Collaborative Structures and Constructware,[13] but similar developments soon followed in the United Kingdom and other countries as the dot.com boom took hold. Capable of being accessed via any computer equipped with a browser and an internet connection, these did not tie end-users to particular networks, and did not compromise corporate network security by allowing outsiders to penetrate firewalls, etc. Moreover, often being provided by a new type of software provider – ASPs[14] – they fitted with industry trends in favour of outsourcing. With an independent organisation being paid specifically to host the system, clients or project team members no longer needed to commit considerable IT resources to the system's acquisition, implementation and support, nor did they require any major investment by other end-users in new hardware or software to access the system (an important factor, perhaps, when partnering with smaller suppliers and other project participants).

With websites, email and intranets already improving internal communications, the challenge was to extend the communication up and down the external supply chain.

2.2.5   The development of the UK construction collaboration market

This challenge was taken up by both established names and new start-up businesses (the range of entrants to this market sector is covered in more detail in Chapter 3). During the dot.com boom, there were optimistic forecasts of exponential market growth, and it was estimated that around 100 different providers were targeting the United Kingdom. These included some established software vendors and some established AEC industry names (e.g. UK contractor Bovis and multi-disciplinary practice Arup). There were several overseas entrants, including some from the United States, while the home-grown contenders included 4Projects, BIW Technologies, BuildOnline,[15] Business Collaborator, Cadweb, ePin, Sarcophagus and Union Square.

No single business model dominated. Some of the collaboration providers could trace their roots back well before the 'dot.com boom'. Others were start-ups heavily backed by venture capitalists and launched during the boom. Some were IT spin-offs of existing AEC businesses or emerged as e-business subsidiaries of larger companies.

Others were offerings from existing vendors of other IT systems. The roots of their technologies varied too: some were modified EDMSs, some were primarily intranet tools, while others were focused in inter-organisation communication. Some were focused on document management, others were stronger in drawing management, while some were extensions of e-procurement tools. But almost all began to market themselves as offering web-based 'collaboration systems' (or 'project extranets'), and during 1999–2001, as the UK construction market began to wake up to the possibilities of online collaboration, press features, websites,[16] conferences and exhibitions were quickly filled by businesses all claiming to be able to deliver construction collaboration technology.

As mentioned, the technology came to prominence during a period of dot.com hysteria to which the construction industry was not immune. When the 'dot.com bubble' burst, several businesses soon merged or disappeared altogether. Coupled with the confusing proliferation of business models and different software applications, the dot.com implosion doubtless fuelled uncertainty about the financial stability and long-term prospects of some of these businesses.

Crucially, though, project extranets had already begun to achieve some credibility with a few leading AEC clients, particularly those who were already adopting partnering approaches. If a client and its supply chain were committed to the transparency and trust afforded by genuine partnering, collaboration technology was a logical next step, and some clients were prepared to experiment despite the technology's lack of track record. For example, in 1999, after testing the system on a store project in Clapham, south London, retailer Sainsbury's – one of the United Kingdom's largest building clients – adopted BIW's system as its corporate standard, heralding a dramatic increase in the use of collaboration technology by its supply chains and, in due course, by those of other prolific client organisations.[17] Looked at in terms of the product life-cycle, such 'innovator' and 'early adopter' customers also included Manchester Airport and Marks & Spencer (BIW), Prudential (Cadweb), Debenham's (ePin), Asda (BIW and Sarcophagus)[18] and Bass Leisure (4Projects). Another committed partnering organisation, airports operator BAA, built on its earlier framework agreement experiences: in early 2002 it was using services from BIW, Bidcom and 4Projects to manage project communications; it later focused on services from the first two but delivered through the Asite portal (launched in 2001).

Alongside partnering, a related stimulus was the shift towards combining finance, design, build and operation activities – e.g. through Private Finance Initiative (PFI) and Public Private Partnership (PPP) projects. Instead of focusing just on design and build activities, the project team – and therefore the collaboration challenge – was enlarged to include organisations responsible for raising funding for the project and others who will be responsible for operating and maintaining the completed facility through a concession period which may last 25 or 30 years. A PFI client would often be a 'special purpose vehicle' or consortium including financial institutions, the owner/operator, contractors and consultants, and between them they would need to compile a huge volume of documentation to complete a successful PFI bid. Collaboration technology was the obvious answer for such consortia both as a bid management tool and to manage the project documentation; the central repository of as-built data then serves both a powerful resource to satisfy the demands of audit bodies and as a key building block for efficient future 'whole life' management of the facility.

In addition to client organisations, various contractors and consultants also began to experiment with collaboration technology. Where prolific clients had already begun to standardise on particular systems, their project teams would normally be mandated to use them; in some instances, this meant businesses working for multiple clients were expected to become proficient in using several different systems. This had three effects: first, as contractors and consultants might be using several different systems in parallel on projects for different clients, training and familiarisation processes were laboriously repeated; second, it created some marketing confusion with different vendors each claiming to have the lion's share of the leading contractors or consultants; and third, it also helped such businesses to test the different applications and identify their own preferred options. Thus, when a typical 'occasional' client might ask a contractor or consultant to recommend an appropriate system, they were better equipped to do so. Indeed, several formed strategic relationships with particular vendors. For example, contractor Balfour Beatty decided to standardise on BuildOnline's system; Kajima and Gleeds announced deals with BIW; while WS Atkins settled on using Business Collaborator as its company-wide solution.

2.2.6 The UK take-up of construction collaboration technologies

Through adoption by major client organisations and through promotion via contractors and consultants, the proportion of UK construction businesses employing collaboration technology began to grow.

A July 2001 e-business survey by the Construction Confederation revealed that just 4 per cent of all 292 contractor respondents 'used the internet for project collaboration',[19] with an additional 10 per cent planning to do so after 2002. Larger firms were more likely to use the internet for collaboration: while there was no use at all among contractors with under 30 employees, 4 per cent of firms with 30–250 staff did, as did 23 per cent among firms with 250–1,000 staff, and 43 per cent of contractors with over 1,000 employees.[20]

A few months later, from a survey of 256 CAD managers, the Business Advantage Group (January 2002) estimated that one in five UK construction sites were operating a web-based project hosting service; again, there were marked differences according to company size: only 10 per cent of sites with less than 25 staff had used such a service (mainly very small architectural practices), compared to 32 per cent of sites with more than 100 staff. A *New Civil Engineer* poll (Hansford 2002) found 38 per cent of respondents said they were using online project collaboration (and forecast – rather optimistically – that 55 per cent of teams working on projects worth more than £5 million would be using such systems in 2003).

Barbour (2002, p. 31) found that, on average, 2 per cent of projects in 2001 were managed using project collaboration tools or extranets, with use greater among larger companies; a year later (Barbour 2003, p. 14), it said 13 per cent of the 322 client respondents to its telephone survey claimed their teams used such technology, rising to one-third of those spending over £100 million.[21] In June 2004, the IT Construction Forum (ITCF) (the relaunched ITCBP Programme) surveyed 373 firms of all

Table 2.1 Number of users of collaboration systems,
           March 2003

Technology provider	Number of users
4Projects	13,000
Bidcom	6,000
BIW Technologies	23,000
BuildOnline	10,000
Business Collaborator	15,000
Cadweb	6,000
Causeway	7,000
Sarcophagus	7,200

Source: Compagnia 2003.

types and sizes, and 43 per cent of its respondents said they used project extranets to collaborate online. The DTI benchmarking study (2004, p. 52) found 17 per cent of construction businesses claiming to be extranet users. Use among small contracting businesses remained low, however. A 2004 survey of more than 800 members of the National Federation of Builders – which represents over 3,000 medium sized contractors and smaller builders throughout England and Wales – found that only 3 per cent of respondents had used an extranet or project collaboration tool.[22]

While it is difficult to get reliable and consistent statistical information, surveys and vendors' own figures on the numbers of users also show that adoption of the technology has been growing. In August 2001, a ConstructionPlus survey of some 20 collaboration providers (cited in Alshawi and Ingirige 2001) estimated there were 25,000 UK users, working on around 1,500 UK projects. Two years later, that number had more than tripled; a 2003 survey by e-business consultancy Compagnia (2003; see Table 2.1) suggested that the eight leading UK providers alone had a total user community of over 87,000, though it did indicate figures might need to be adjusted to account for individuals using more than one system.

Individual providers vary in their willingness to release statistics about their user base. However, from 3,400 users from 570 companies at the end of 2000, usage of BIW's system grew to 40,000 by the end of 2004. At the same date, 4Projects claimed to have 36,000 users. In other words, between them, these two vendors claimed to have added a further 39,000 users in the 21 months since Compagnia collated its figures. Allowing for some overlap between user groups, and accepting there has also been modest increases in their competitors' user communities, it is probably fair to say that by the end of 2004 over 150,000 industry professionals had experience of using one or more of the leading eight UK products. Adding the users of other vendors' products and of in-house developed systems, the total user community for web-based collaboration systems in early 2005, was probably somewhere around 175,000.

Partly because it had begun to achieve credibility with some leading clients and partly because take-up was beginning to achieve critical mass, online collaboration has started to become an accepted part of AEC life, progressing from the 'novel' to the 'normal' (in many teams, particularly those embracing partnering-type

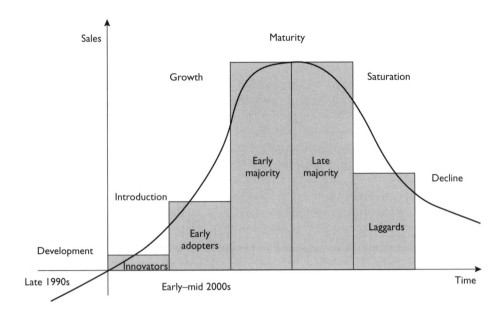

Figure 2.1 The product life cycle.

approaches, the question posed was no longer 'should we use the technology?' but 'which technology should we use?'). Analysed in terms of the traditional product life cycle (see Figure 2.1), it could be argued that the technology had successfully negotiated the 'development' and 'introduction' phases, and by the mid-2000s had embarked upon the 'growth' phase. Awareness and interest was stimulated by 'innovator' clients; 'early adopters' then started to join in; the 'early majority' is, perhaps, just beginning to embrace the technology.

2.3   The convergence of culture and technology

To an industry that was – and to some extent still is – synonymous with delay, waste and inefficiency, there are now well-documented paths to help businesses develop genuine partnering, with both industry and government encouraging and facilitating such approaches. Technological innovation now provides a powerful communication platform to support and enhance that partnering effort, and, as we have seen, clients and their project teams are increasingly being exhorted to integrate their IT systems. Construction collaboration technologies emerged at the right time to fill a need and meet the aspirations of the more progressive AEC industry clients and their supply chain partners, and were able to establish some critical early credibility within a notoriously sceptical industry.

However, the technologies' movement towards the UK industry mainstream has been slowed by at least two factors. First, the 2000 dot.com implosion fuelled uncertainty about e-businesses' financial stability and long-term prospects, and provided a convenient excuse for some to delay decisions about using them (alongside sometimes spurious issues such as legal admissibility of electronic documents, absence of industry standards, additional telecommunications, hardware and software requirements, data-centre locations, internet connectivity, security concerns, and so on – all addressed in the following chapters).

Second, partnering has yet to become the dominant approach to delivering construction projects. As mentioned towards the end of Chapter 1, industry estimates suggest it is only embraced on around a third of all projects. Similarly, there is evidence (e.g. in Green *et al.* 2004) that the AEC industry's attitudes to supply chain management vary: those involved in prime contracting saw it 'as an essential means of ensuring their competitive position.... Others were openly sceptical of what they saw to be the latest "fad"' (p. 35). As a result, demand for those technologies that allow higher levels of communication and information transparency may only grow as the enthusiasm for partnering grows. Alternatively, the applications may need to be configured or customised to mimic traditional project information controls and so allow different project team members to electronically restrict access by some team members to certain types of information. There may still be some benefits from using the technologies in this way – Chapter 1 described them as 'construction communication technologies' – but the gains will, arguably, be much more modest. Such teams might achieve one–off savings on tangible costs such as postage and printing, but will miss out on the significant efficiency improvements arising from adopting more integrated approaches (as mentioned, the RCF envisaged potential savings of 30 per cent).

For genuine collaborative working to succeed, there have to be significant changes to both the culture of the teams involved *and* to the tools they use to manage their communications. As Sir Michael Latham told *Public Sector & Local Government* magazine in January 2002: 'Apart from the obvious advantages of avoiding confusion, duplication of effort and waste, improved communication also lies at the heart of the modern trend away from the adversarial system, towards partnering and collaborative working.'

2.4   Chapter summary

This chapter has presented a necessarily brief overview of the recent history of partnering and of the emergence of collaboration technologies within the UK construction industry. Since the dot.com bubble burst in 2000, industry professionals have begun to learn about the technology, even though, in many respects, the functionality of the software should not be the first or even the most important consideration. Customers need to have a detailed understanding of the backgrounds of the vendors and the business models the vendors have adopted. Particularly as the tools will be used to manage multi-million projects of long duration, customers will need to decide whether they wish to manage the technology themselves or have someone else manage it on their behalves. Only then perhaps should the features of the software applications themselves, and their perceived benefits, begin to become a factor.

The Chapters 3 and 4 are devoted to developing an understanding of the key differences between the various vendors of construction collaboration technologies (some already mentioned), and to learning more about hosting and the ASP model.

Chapter 5 will focus on software functionality and service support issues. Chapters 6 and 7 will deal with internet connectivity and with the legal issues respectively. Building on this chapter's consideration of partnering, Chapter 8 will bring us back to a consideration of the people and process issues critical to successful selection and introduction of the technologies into project team environments. Chapter 9 will analyse the benefits claimed for use of the technologies.

Chapter 3

The construction collaboration providers

This chapter:

- will help the reader understand the main generic differences between the leading collaboration technology vendors with a UK presence;
- describes the history of the main players and what the key selection factors might be;
- provides an overview of software licensing practices.

As was outlined in Chapter 2, the early history of construction collaboration technology in the UK AEC market saw a host of companies burst upon the scene during the late 1990s and early 2000s. These included some familiar software vendors, some established construction businesses and several new start-up businesses.

At the heights of the dot.com boom, some commentators estimated that over 100 different collaboration technology providers were targeting the UK market, and more than 200 worldwide. But, with the benefit of hindsight, these were overestimates. For example, some figures were based on the number of vendors listed on various extranet websites, but some of these sites (and therefore many of their listed vendors) were not focused solely on the AEC market, while some of the vendors' products were not aimed at collaboration *per se*, or were often small applications developed to improve particular tasks or processes.[1] Particularly where information was drawn mainly from US-based sources, it featured many vendors who were not operating outside the United States. And even when United States or other overseas ventures did say they were intending to do business in the United Kingdom, this may well have been ambitious dot.com hype intended to reassure actual or would-be investors and customers about the venture's ambitious plans for international expansion (it is worth recalling, also, that the dot.com boom also saw talk of 'vapour-ware' – solutions which had yet to be developed, or acquired, but which dot.com businesses were already marketing as though they would be immediately available).

Now that the dust of the dot.com crash has begun to settle, we can begin to separate the hype from the reality – for a start, the total number of collaboration vendors active in the UK AEC marketplace is probably a fifth of the original estimate – and to analyse the early history of the construction collaboration technology marketplace that actually did develop in the United Kingdom. Armed with this information, readers should then be in a better position to make informed choices about which (if any) of the vendors and solutions might be most suitable for their organisation, project or programme of works.

Particularly as the technologies (and many of the vendors) are still regarded as relatively new and untested, several commentators have stressed that buyers should look beyond the functionality, 'look and feel', and cost of the applications offered; instead, they suggest, buyers should play close attention to the vendors themselves. This and the ensuing chapters will look at each of the main factors in the buying decision. This chapter takes a snapshot of the scene that had emerged by 2005; looking at most of the leading vendors with a UK presence, it seeks to help readers understand the main differences between them. In particular, it gives a quick overview of the history behind the main players, their experience of the UK market, the origins and development of their software, and their licensing/charging structures.[2]

3.1   Understanding the providers' roots

For the purposes of this chapter, the UK construction collaboration technology market can be broken down into five main groups (each is capable of further subdivision, and there is some overlap between some of the categories, but for structural simplicity we will stick with five):

- traditional software vendors and resellers, including those selling both generic and AEC-specific applications;
- traditional AEC businesses, who developed their own applications or bought third-party ones;
- UK-based construction collaboration software vendors;
- non-UK-based construction collaboration software vendors;
- bespoke in-house systems developed by some client organisations.

3.1.1   Traditional software vendors

When we attempted a definition of collaboration technology in Chapter 1 it was clear that, depending on one's point of view, many different applications – some of them from well-established software vendors (both inside and outside the AEC industry) – could be described as collaboration tools. Generic applications such as intranet or enterprise portal solutions, groupware, EDMSs, file-sharing, and knowledge management tools could all be applied by organisations involved with construction projects. These might be suitable in other industries, especially if project interactions were primarily document-based and confined to single organisations. However, the extensive requirement to create, exchange, view and edit large and complex drawing files can impose particular burdens, as can the need to do so across short-lived, multi-disciplinary, multi-company, multi-location project teams. Moreover, AEC teams tend to be comprised, more often than not, by people who can be techno-phobic, working on limited budgets, with only limited IT and telecommunications resources at their disposal. Even if the infrastructure issues could be overcome, such generic EDMS applications would often also need extensive customisation and/or training to be of value to the end-users.[3] For large client organisations, particularly those managing a steady stream of major capital projects and/or working with a fairly stable supply chain of contractors, consultants and suppliers, customising an EDMS or enterprise portal application (e.g. Documentum's eRooms, Hummingbird's DOCSFusion) or augmenting an IBM Lotus implementation (e.g. using Lotus Team Workplace)[4] might well be viable. Otherwise, the software vendors might seek to enter the market through alliances with specialist AEC providers; for instance, OpenText's web-based collaborative application Livelink was developed into an offering by UK-based AEC IT vendor, Causeway Technologies (see later) and into Canadian eBuild.ca's collaboration solution.

Of course, some global vendors of traditional software were already familiar with the needs, and the low profit margins, of the AEC sector; they knew what type of functionality was required; their business models were adapted to this low margin industry; and they also tended to have existing relationships with clients in the industry and so could generate contacts, and in due course contracts, with less marketing expenditure on building awareness. For example, US CAD software giants Autodesk and Bentley and project management tools providers Meridian Project Systems and Primavera Systems quickly realised that they could add collaborative functionality to their product portfolios. Bentley developed ProjectWise and Viecon (both applications designed to be customer-hosted). Meridian launched its ProjectTalk service in June 2000, while Primavera followed suit with PrimeContract in November 2000 and its locally hosted Expedition product. Within the UK construction market, perhaps only ProjectWise and Buzzsaw emerged from the major AEC software names as a player in the mainstream collaboration market:

- Bentley's ProjectWise saw some take-up among architects and other consultants (e.g. at London architect Damond Lock Grabowski – see Building Centre Trust 2000b), buoyed by its existing footprint within the MicroStation CAD user community.
- Buzzsaw – in October 1999, Autodesk was a 40 per cent shareholder in its spin-off Buzzsaw.com, which launched its ProjectPoint online project collaboration service soon after, having raised around $90 million before the dot.com crash. After Buzzsaw posted an operating loss of over $50 million in 2000 and began laying people off in May 2001, Autodesk acquired the business in July 2001 and ProjectPoint was later re-branded as an Autodesk product: Buzzsaw.

Within the United Kingdom, some smaller AEC software vendors also sought to offer collaboration applications that could potentially be integrated with their existing software products (e.g. ECL's Information Manager, and Sysnet's Sysdox). The main player to emerge from this direction was Causeway Technologies, a vendor of site management and estimating packages:

- Causeway – in August 2000, Causeway announced that it would be incorporating OpenText's Livelink into its buildingwork.com e-construction portal; in December 2000, AEC consultancy Arup joined its e-construction alliance – a step cemented in May 2001, when Arup's in-house developed product, Integration, was acquired. The business eventually launched Causeway Collaboration in the United Kingdom in November 2001.

3.1.2   Traditional AEC businesses

To augment, differentiate or improve their core business offerings of design, project management or contracting, many consultancy and contractor firms have developed their own software tools (indeed, some firms have created viable sidelines in software development – Oasys, for example, is the software arm of multi-disciplinary consultancy Arup). Often these would be created to tackle particular tasks (e.g. modelling water quality in tidal estuaries, predicting ground subsidence above tunnelling works, or modelling people movement through airports or railway stations), but, particularly where firms had extensive project or construction management responsibilities, thought quickly turned to developing in-house applications that might help manage the mass of drawings and documentation involved – applications that could potentially also differentiate their providers to customers.

In the United Kingdom, contractor Bovis developed Hummingbird[5] Image Management, a document management system[6] employed on most European Bovis schemes including the Bluewater shopping centre development in north Kent (Building Centre Trust 2000a). Multi-disciplinary consultancy Arup developed its own free Windows-based file navigation and viewing software, Columbus, and the aforementioned Integration. And construction and project manager Schal (part of contractor group Carillion) developed Project Information Management System (PIMS) and employed it on the redevelopment of London's Royal Opera House.

Specialist subcontractors also sought to differentiate themselves by offering information management services. For example, retail fit-out contractor Styles & Wood developed StoreData, and building services specialist Commtech Group was considering an extranet service to enhance access to building operation and maintenance information. Printing services business ServicePoint developed ProjectVault not so much as a collaboration tool but more as a secure online data repository to complement its core printing and scanning services.

If in-house software development resources were unavailable, then an AEC consultancy could always look at buying a third-party application and developing a branded solution based on that. For example, Gibb (now part of the US-based Jacobs Engineering Group) used ActiveProject from US-based Framework Technologies; WS Atkins' iProNET (internet-based Project NETwork), launched in 1999, is based on Business Collaborator (see Section 3.1.3); and Rock Consulting's E-box offering is based on the Citadon product, ProjectNet (see Section 3.1.4).

3.1.3   UK-based construction collaboration software vendors

Existing AEC businesses and AEC IT vendors (e.g. Autodesk, Bentley, Ramesys, COINS, etc.) developing their own solutions had a key advantage in the early days of the construction collaboration technology market: they were generally familiar names and needed to do little to raise their profile or establish their industry credibility. However, in the United Kingdom most failed to capitalise upon this advantage. Instead, they were soon outnumbered by an array of new entrants to the market, both UK-based and overseas. Some were internet 'pure-plays', not all of whom actually had applications to offer;[7] others were established industry players looking to find new routes to market or looking to offer new products.

Most of the UK-based candidates were formed during the mid-late 1990s, often from within existing AEC businesses, and came to prominence as the dot.com boom peaked. Some of these businesses grew organically from small beginnings; others attracted venture capital backing to build their infrastructures and market share more quickly. They included (in alphabetical order):

- 4Projects – started as joint venture between publishing group Leighton and construction business Taylor Woodrow and began delivering services in 1998, becoming a wholly owned subsidiary of Leighton in 2000.
- Architec – a small architect-led business focused on architects' projects, Architec relocated from Canada to London in 1998.
- Asite – a comparative late-comer to the UK scene but with some powerful backers, the Asite construction portal was formed in 2001 and initially relied on re-selling collaboration products from other vendors (ProjectNet from Bidcom Ltd and BIW Information Channel from BIW). However, in 2004, it added its own collaboration software (Asite Project Workspace and Asite Project Workflow) to its portfolio of tools, ceased reselling the Bidcom and BIW solutions, and announced it would also be selling the German company AEC/communications GmbH's collaboration product 'think project!' in the United Kingdom, branded as 'Asite Enterprise Workflow'.
- BIW Technologies – BIW was founded in 1994 as part of a government-backed construction portal project. It began delivering collaboration solutions in 1998, and, under a new management team drawn from construction IT backgrounds, was re-launched as an independent business in January 2000 with BIW Information Channel as its core product.
- BuildOnline – started in 1999 in Ireland by property entrepreneur Brian Moran to create a Europe-wide online market for the construction industry, supplying building materials and services (TradeOnline and SuppliersOnline), before relocating to London and refocusing its efforts on collaboration (ProjectsOnline, rebranded in November 2004 as BuildOnline Collaboration On Demand) and tendering solutions. Also operates businesses in Germany and France.
- Business Collaborator – established in Reading in 1997 and was part of software solutions arm of the Enviros environmental consulting and software group until acquired by CodaSciSys in April 2003.
- Cadweb – formed in 1995, Cadweb launched its first commercial product in 1998, with its Cadweb.net solution launched in March 2002.
- e-Hub.com – formed in 2000 and based in Northern Ireland, e-Hub.com developed project-hub.net, but it achieved only limited take-up, and the company now concentrates on e-procurement services for the Irish market.
- ePIN – developed as a privately owned, privately funded company from within an established engineering consultancy, and launched its ePIN Online Data Manager on live projects in 1998.
- RS (UK) – established in 1989 in England with a background from the offshore petrochemical industry and technology, and focused on document and drawing management.

- Sarcophagus – started in the early 1990s as part of a UK multi-disciplinary design company, based in West Yorkshire, separating in August 1999 to focus on providing its project extranet service, the-project.co.uk.
- Union Square Software – a privately owned business focused on customer-hosted knowledge management and enterprise portal solutions, notably Workspace.

3.1.4 Non-UK-based construction collaboration software vendors

Given that the United States is, first, home to many of the world's largest software firms, and, second, one of the world's largest AEC markets, it is little wonder that construction collaboration technology developed quickly in the United States, with some of the earliest products appearing in the mid-1990s. Particularly during the dot.com boom, many North American vendors began to look beyond their own shores for the next big opportunity, and, given that it was another English-speaking market, the United Kingdom was an obvious target for many of them. Accordingly, during the very early 2000s, various North American businesses began to investigate the UK market,[8] including Cephren (formerly BluelineOnline and eBricks), Bidcom Inc. (incorporating Cubus, it merged in March 2001 with Cephren to become Citadon), e-Project Enterprise, Hummingbird, Meridian Project Systems, ProjectVillage and Web4.

Other overseas providers[9] to look at the UK market included Israel-based I-Scraper,[10] Germany's AEC/communications GmbH, Belgium-based Bricsnet, France's Constructeo (since merged with Bricsnet), Swedish start-up ProjectPlace, and Australian providers Aconex and QA Software (provider of TeamBinder).

Of these non-UK-based construction collaboration software vendors, only Aconex, Bidcom Ltd and, to a lesser extent, QA's TeamBinder managed to build any significant UK AEC market presence:[11]

- Aconex – launched in Australia in 2000 to offer both collaboration and procurement management services (i.e. Australian Construction Exchange), Aconex's UK presence was focused on collaboration with its core product AconexAEC. Its early UK successes included work for contractor Bluestone and a deal with Australian-based contractor Multiplex.
- Bidcom Ltd – established in the United Kingdom in 2000 by its American parent (Bidcom Inc., later Citadon), Bidcom Ltd was initially backed by KPMG and three well-known AEC businesses: Carillion, Wates and EC Harris. ProjectNet was among the products initially used by airports operator BAA. However, by late 2004, its original backers had retreated, it was no longer a Citadon company, and the ProjectNet product was being resold in the United Kingdom by 'E-box', a subsidiary of Rock Consulting, a west London construction consultancy whose expertise included PFI process management.
- QA – Melbourne-based QA Software have developed solutions for document management (e.g. QDMS) and correspondence management (QTRAK). Its web-based collaboration solutions include TeamBinder and, for intra-enterprise use, WebDocsPro.

3.1.5   Bespoke in-house systems developed by, or for, some client organisations

While not strictly offered as products on the open market, some client organisations had no immediate need for any of the systems identified so far, having already followed a similar route to some AEC businesses and either developed their own bespoke solutions, or bought in a third party solution, to suit their particular needs. For example, pharmaceutical firm GlaxoSmithKline commissioned its own LAN-based, on-site EDMS/intranet (Technical Document Management (TDM) from former Bovis man Ray Crotty's C3 Systems) for use by some 200 different contractors and consultants engaged on its west London headquarters project in 1998; UK property giant Canary Wharf commissioned a third party to develop bespoke IT systems to support its construction and development plans.

3.2   History, management and financial status of technology providers

As should be clear from the preceding overview, it is difficult to identify any common traits shared by all the leading providers operating in the UK market. They have different origins, have been financed in different ways, and have developed in different directions. However, given that customers will be relying upon them to deliver and – in many instances – to host and maintain collaboration technologies that will be used to run key capital asset delivery projects worth many millions of pounds, it is worth devoting time and effort to discovering and understanding more about the history, current management and financial stability of a prospective provider.

3.2.1   Financial stability

Given the relative immaturity of the market, lawyers and others have urged would-be customers to take precautions in case a chosen vendor becomes insolvent (see Chapter 7 for a more detailed discussion). Prevention is better than cure, and customers contemplating a lengthy commitment to a business-critical relationship should seek detailed information about the financial status of a provider (e.g. sources of funding, audited accounts, management accounts, shareholder details, insurance cover, etc.) before entering into a contract. Clearly, it is important to verify that the provider is financially secure and not about to change hands or go bust – should such an event occur during a project, the impact on that project and on all project participants, could be severe.

However, it is also worth looking at the past history of the business. For example, a software vendor that has grown organically will have different financial pressures to one that has been heavily funded by venture capitalists, while a business that started out as a traditional software vendor will have different pressures to one that started out as an ASP. If a vendor has changed owners, the customer should look at the reasons behind the sale and perhaps make enquiries about the levels of support guaranteed by the current parent company.

Where vendors offer collaboration solutions from among a portfolio of other products or services (e.g. Causeway), the buyer will, of course, need to assess the extent to which a vendor is committed to supporting that particular application. This applies particularly if it is offered as an ASP solution while all of the vendor's other products are delivered as traditional client–server software packages, or if the IT solution sits alongside a range of conventional AEC contracting, consulting or other services.[12] And the financial assessment of that vendor may require the buyer to ask what proportion of the vendor's revenues and profits (if any) come from collaboration, while also checking whether information on funding, numbers of customers, staff numbers, user numbers, etc. relate to the whole product/service portfolio or just the collaboration offering.[13] Similarly, if the vendor serves more than one major industry sector or is active in several geographical regions, it is vital to understand what proportion of revenues and resources are associated with construction-related collaboration and/or UK-based projects in that sector. A healthy group profit may be used, for example, to disguise losses in a particular market or in relation to a particular product or service.

3.2.2 Relevant experience

Particularly in the early days of the UK market, customers were often advised to ensure that their chosen vendor had experience of working on UK construction projects. This was, of course, partly due to the arrival in the United Kingdom of several overseas businesses whose track record (if any) was gained in other markets (e.g. United States)[14] where construction processes are markedly different. From formal contracts and procedures to less formal, but nonetheless important, processes employed to manage day-to-day team interaction, projects are managed in different ways in different countries. Regardless of how extensively the software can be customised or re-configured, project teams need to be sure that the vendor's staff involved with implementation and support (e.g. helpdesk) are familiar with how projects are delivered in their local market.

To help gauge the depth of experience, prospective customers should ask vendors about how many UK customers they have (and, ideally, should seek names of individuals who can provide independent references about the vendor, its system and its implementation, training and support services), and how many AEC projects and end-users they have. In reviewing a client list, prospective customers should also find out how many are *current* clients (lists may include organisations who tested the vendor's system on one pilot scheme in the past and did not use the system again); they should also check which are the paying customers (lists might include consultants or contractors who are end-users but who had little or no influence over the choice of system used by the paying customer).

It is also worth finding out about the origins of the business to understand more about the experience and expertise of the vendor's management team and its support staff. Do they have experience in selling industry-strength IT systems within the UK AEC sector? Has the business always been focused on collaboration software? If not, what other systems did/does it offer? How long has the current management team been in place, and where were they before they joined the collaboration vendor? Are the vendor's consultants IT specialists or construction professionals (or a mixture of both)? It may also be advisable for key project team members to meet the vendor's consultants so that they can ascertain their technical strengths and weaknesses and evaluate the all-important 'chemistry' that will help build a successful relationship between the vendor and the end-users.

3.2.3   Software history

Linked with the above point, but also striking quite fundamentally at the origins of each technology provider, is where and how the software has been developed. Does it have a coherent architecture? Or is it a collection of disparate pieces of code that may be difficult to adapt as customer requirements change?

- *United Kingdom or overseas software?* The roots of some systems offered in the United Kingdom lie overseas and/or in generic systems; they may therefore have been adapted for use in the United Kingdom and in its AEC industry in particular. Would-be customers need to ensure the vendor's software development team has detailed knowledge of all parts of the application.
- *Client–server based, web-enabled or 'pure ASP'?* The different software architectures can radically affect their implementation, use and future development (see Chapter 4 for a much more detailed analysis of the pros and cons of the ASP model).
- *Single or multi-authored?* If the vendor's business has gone through a number of mergers or acquisitions, the resulting software may be a mélange of different packages. Even if there have been few corporate changes, it may be worth checking how long key developers have been with the vendor. Again, customers need to check that the software development team retains detailed knowledge of all parts of the current application.
- *UK-based or overseas software development?* Some vendors have moved software development overseas, for example, to India. Customers should ask about how changes to the software are communicated, and how quickly these are delivered. Can customers or project team members brief software developers face-to-face about their requirements? How responsive is the vendor's software development team?
- *Own IPR or licensed?* If a vendor's application is based on a third party's software, or if key areas of functionality (e.g. drawing viewing, mark-up and commenting technology) require third-party plug-ins, then the buyer should enquire about how these technologies are licensed. Will the buyer need to pay additional license fees to the third party to use their applications, or will the collaboration technology vendor be absorbing these costs as part of the price for its services? What undertakings has the vendor received from the third party about continued compatibility or integration between the different packages? If the vendor is absorbing the third party costs, how do these expenses impact on its business model, particularly if the third party suddenly increases its license fees?
- *Long-term 'road-map'* Customers will be aware of how fast-changing the IT world is, and will need to talk to vendors about their future development plans for the software. Hardware and software architectures and operating systems continue to develop, and the ability of IT to support previously untouched corporate needs continues to grow. Has the vendor got a clear and realistic vision of how its application will develop over the next few years? How far does this vision coincide with the customer's own potential business and/or IT development plans? What influence (if any) will the customer have over the future development of the application?

3.2.4 Industry independence of provider

In the early days of e-business in the United Kingdom (as discussed in Chapter 2), there was considerable industry interest in establishing AEC e-business portals (e.g. Mercadium, Arrideo, etc.), most of which withered quickly as the dot.com bubble burst and e-business hype was replaced by a more sober recognition of what could actually be achieved in the foreseeable future. Construction collaboration emerged as the most viable offering, and some providers attracted investment from major AEC businesses (e.g. Bidcom Ltd was backed by two UK contractors and a major consultancy; BuildOnline forged strong links with contractor group Balfour Beatty; and Asite's backers included several major owners and developers including AXA, BAA, British Land, Stanhope, Prudential, Legal & General and – its biggest shareholder – Rotch). However, this threw up further issues.

From the vendors' point of view the backing of a well-known industry name might help convince some procrastinators that the provider was more stable and that the software had been tried and tested by that backer.[15] On the other hand, just as some buyers were wary of using e-marketplaces lest their purchasing data be accessed by competitors, so some would-be collaborators could be wary of using systems provided by a direct competitor, or of using industry-backed systems because they feared their design data, construction methodologies, tender documentation, etc. might be viewed by competitors who were shareholders in the solution provider. Similarly, a strong concentration of major clients (as in Asite, e.g.) could excite concerns that consultants, contractors and suppliers could be mandated to use a particular collaboration platform as a condition of project participation regardless of whether it was, in fact, the best solution for the project or the preferred choice of the project team, with the clients benefiting doubly: through lower project team reimbursable overheads and through dividend payments to them as shareholders.

Clients or asset owners may also have reason to be wary of the provision of information management systems by supply chain members such as a specialist subcontractor. If they focus on particular markets or types of information, such providers will not be gathering lessons learnt from other client types or from managing different processes. Moreover, their IT offer will not usually be part of their core business – subcontracting, for example, can be very volatile, prone to downturns, litigation, etc.; potential customers will need reassurance that the IT service will continue even if the main business experiences problems. There is also a need to maintain appropriate commercial relationships and high levels of information confidentiality. A prolific client may wish to use more than one supplier; it may not want one supply chain member to hold data on all its projects; and, even if the client permitted such a practice, would a rival supplier by confident that its data would remain confidential on a system hosted by a competitor? To retain flexibility in managing supply chain relationships, the ultimate customer may prefer data to be held by an independent third party, perhaps one solely dedicated to delivery and long-term support of IT solutions.

3.2.5 Software 'lock-in' or industry standard?

The extent to which a vendor's software conforms to industry standards is also a major issue. If undertaking several projects in parallel for several different clients – each of whom mandates the use of a particular system – a contractor or a consultant, for example, may find themselves having to use more than one construction collaboration platform. If the different applications require the same basic hardware and software, have common features, adopt similar document naming, numbering and revision control conventions, are intuitive to use and are broadly similar in 'look and feel', then switching between the different applications can be relatively straightforward. However, if they have different hardware and software requirements, work in different ways, using different nomenclature, the end-user may need to be retrained so that he/she can migrate from one solution to another, while back-office processes such as drawing numbering systems may need to be amended to suit each project/system.

Thankfully, several of the main providers in the United Kingdom have begun to look at standards issues. The Network of Construction Collaboration Technology Providers was established in 2003 partly to devise interoperability standards to which the member providers must comply. Such standards would help ensure that all information could be easily transferred should an ongoing project need to be switched to another system (e.g. if the vendor went bust); a longer term objective was to devise a standard that would allow an architect, for example, to adopt one application as its preferred choice and use its interface to access information stored on any of the NCCTP members' systems.

3.3   Charging structure

Related to the issue of financial stability (see Section 3.2.1) is how the vendor requires its customers to pay for its collaboration solution. As the market can be roughly divided into vendors of traditional, locally hosted software and those offering ASP solutions (and some offering both), this is not a straightforward issue. It is therefore worth looking in more detail at the whole issue of software licensing.

3.3.1   Overview of software licensing

Computer software is one of the few industries in which a customer buys something but does not actually own it. In most instances, the largest proportion of the price of a software application relates to the cost of the licence to use the software and its related data, not the costs involved in physically packaging, delivering, documenting and supporting it. Instead, the software purchase mainly involves acquiring a licence to use a product that remains the intellectual property of the software vendor who imposes restrictions on that product's use by the customer and/or the end users.

Because different customers have varying needs, vendors have been able to offer different forms of licences according to their customers' requirements. Traditional software licences tend to have two dimensions: the time period covered by the licence, and the number of users (often licences combine elements of both):

- Time period-based licences, broadly, cover two categories: 'perpetual' licences, which have no limitation on the period of usage (off-the-shelf packaged software products typically have a single user, perpetual licence normally relating to the purchased version only), and 'term' licences, which have a limited period of validity, typically 12 or 24 months; perpetual licences can be likened to an outright purchase while the term licence is more like a lease.

- User-based licences can be applied, broadly, in two ways: 'per seat' or 'per server' (or a combination of both). In the latter instance, for example, the price of a server licence will be based on an estimate of the number of users dedicated to a particular server.

The emergence and growth of networked PCs prompted some evolution so far as software licensing was concerned, allowing vendors to licence their applications' use in multi-user, shared-computing environments. Multi-user variations include:

- 'concurrent use' or 'floating network' licences, which limit the number of simultaneous users on a particular network, and are typically managed by a logging-in process;
- 'site' licences to cover all users at a particular location;
- 'enterprise' licences to cover all of an organisation's locations, with pricing dependent on the number of servers, frequency of use or number of concurrent users (or combinations of these).

Server-based, network, site and enterprise licences may also have a time dimension. In addition to the initial licence payment covering use of the system for a period of, say three or five years, the customer may also be required to pay additional annual maintenance and service charges (levied as a percentage of the initial licence payment, charges in the range of 15–20 per cent are common). Generally, the vendor of traditional software will tend to seek to limit the licence in some way in the hope of being able to impose additional charges if and when things change.

3.3.2 ASP software licensing

In essence, ASP software is rented. Instead of paying a large, up-front licence fee, customers pay recurring, usually monthly or quarterly fees or subscriptions to use the ASPs software,[16] for as long as they need to use it (in this respect, it is similar to renting a satellite or cable TV service, or using a mobile telephone on a pay-as-you go basis, for example). Moreover, the ASP normally maintains, supports and upgrades the software itself at its own data centres, without little or no input required from, and therefore at less risk to, its customers.

While traditional software sales involve the vendor getting a substantial up-front payment, delivering the software and then, in many instances, getting paid, say, a further 15–20 per cent per annum for maintenance and support, the ASP subscription model means the vendor receives a regular, flow of smaller payments.[17]

3.3.3 Single large payment or regular smaller payments?

The main UK construction collaboration technology providers vary in the payment models they offer their customers. When selecting a provider, customers will need to think about whether they would prefer to pay for their software in one go, or whether they might prefer the rental model.[18] As well as the impact on their own finances, customers should also consider how the chosen software licensing model, and the volume and quality of the deals done, impacts on the financial stability of the vendor.

Vendors who deliver their software on the basis that it will be customer-hosted will tend to adopt the traditional server-based or enterprise licence model whereby the customer pays an up-front free for either a perpetual or a term licence (plus perhaps additional annual maintenance fees for an agreed time period). In some instances, the licence will incorporate a maximum number of users beyond which the customer may have to pay for additional seats. Purchasing a perpetual licence at the outset can prove expensive: for example, a 100-user perpetual licence at £250/user for outright purchase would cost £25,000, with an annual support package adding a further 15 per cent per annum (including inflation, a five-year package might cost the customer a further £20,000). The customer should also be aware that in selling its licence the supplier will have passed a substantial risk to the customer (there can be little guarantee that the vendor will continue trading and/or supporting the software[19]). And, of course, if the customer purchases a term licence, the financial and practical costs of upgrading or replacing the software upon expiry of the licence will also need to be factored into the organisation's future IT planning.

For vendors, large up-front licence payments can make a big and positive impact on the company's finances (with, as mentioned, risk also being passed to the customer). When assessing the vendor's financial stability, therefore, a 'snapshot' taken during a successful period may be misleading as recent performance is no guarantee that the vendor will continue to win work at the same frequency and magnitude. While the vendor can bank the initial payment, there may be little further revenue to follow.

Other vendors provide their software on a pure ASP basis. Here, the customer pays a standard amount per month or per quarter to licence the software for the duration of the project or programme of work. As such vendors will be keen to continue receiving regular payments, it is in their interest to maintain a high quality service, both in terms of the functionality of the software and in terms of the quality of the hosting and support services provided. In this scenario, customers' risks are reduced; as they have not made a large up-front payment, they are less exposed if the vendor goes into liquidation or if they decide to use a different system.

Usually the precise cost is negotiable, but it will tend to reflect the size, value and/or scope of the capital project or programme concerned, as these will help indicate the number of users, the volume of information to be exchanged and stored, the extent of functionality required, etc. Notwithstanding this, the monthly subscription would normally be a fixed amount regardless of the actual number of users or drawings or amount of storage capacity ultimately used.[20] As a payment model, such a project- or programme-related approach is perhaps more likely to encourage collaboration (the cost impact is discussed further in Chapters 8 and 9).

Some vendors seek to licence their application on a per-user or per-seat basis, but this can have the effect, particularly in a cost-conscious industry like construction, of limiting the extent to which the technology is rolled-out to all employees or all members of a project team, as can seeking payment from the end-user companies instead of from the ultimate client or a key member of the core project team. Similarly, payment according to the volume of communications or storage requirements[21] may deter some from making full use of the systems. The main risk here is that there will be only partial take-up of the technology, fragmented and inefficient communication processes, and little or no real collaboration.

Of course, particularly in the early years of the UK collaboration technology market and in the early stages of their vendor/customer relationships, some vendors were keen to lock-in customers to using their technology and would offer quite substantial discounts – at one stage during the dot.com 'land-grab', Buzzsaw.com was even offering its services free – but there comes a point when vendors have to start charging those customers at sustainable rates or risk going out of business.

The cost-conscious approach of many UK construction industry buyers of collaboration technology often meant that decisions were taken almost solely on the grounds of cost. This had several effects. First, it made the vendors of the less expensive solutions initially look more successful, but some customers subsequently found that the performance of the solution and/or the reliability of the vendor's service tended to reflect the lower price paid (infrastructure is expensive and achieving resilience usually means doubling it up), and once a project was finished began to look for alternative suppliers. Second, and moving on from the first point, it meant that some customers either never settled on a permanent solution – in effect, they carried out a series of pilot projects – or only did so after trying several different systems. Third, poor initial experiences with cheaper solutions threatened to fatally damage the credibility of collaboration technology vendors as a whole in the eyes of some customers. Finally, cost-based buying decisions have perhaps delayed rationalisation of the UK market-place.

3.4   The UK construction collaboration technology market

Many, if not most UK construction businesses work on very thin profit margins, and pressure to drive down prices can have an adverse effect on many of the companies with whom they do business. The AEC industry also sees a steady stream of mergers, acquisitions and liquidations – all of which impact on existing supply chains. Similarly, IT firms face constant pressures from customers, investors, competitors and from the pace of technological change, and must continually adapt to maintain their market positions, while the IT industry also sees a constant flow of company mergers, acquisitions and liquidations.

Taking the two market sectors together and focusing on construction collaboration technology in particular, it is also worth noting that the AEC industry was characteristically slow in adopting the new IT tools when they first became available. As we have seen, much of the early adoption was driven by 'innovator' client organisations, with some forward-looking consultants and contractors following quickly in their footsteps. But many other industry customers, consultants, contractors and other suppliers adopted 'one-step-at-a-time' or 'wait-and-see' approaches. In its survey of contractors, Stratagem/DTI (2003) found: 'In many cases, companies are playing the waiting game. They are not pro-actively investing in new technology and are only doing so when a client requests a project collaboration tool'; major businesses were 'reluctant to break out and be the first mover'; and there was 'a certain amount of distrust in new e-business solutions' (p. 23).

IT analyst group Gartner tracks new technologies through a five-stage 'hype cycle': from the peak of inflated expectation, through the trough of disillusionment, up the slope of enlightenment (as the technology overcomes initial problems) to the plateau of productivity. This model was particularly popular to explain the dot.com e-business phenomenon of the late 1990s/early 2000s, and – with the addition of various critical dates from the UK construction collaboration technologies market's timeline to illustrate the different stages – it can also help us to understand where we are today (see Figure 3.1).

Arguably, the AEC industry in the United Kingdom is just beginning to emerge from the trough of disillusionment and starting to make its way up the slope of enlightenment so far as online collaboration systems are concerned. Many AEC businesses have yet to embrace e-business applications as a routine part of everyday commercial life, and the aforementioned reluctance, distrust and conservatism are likely to continue to characterise the adoption process.

For example, a client might decide to test the use of one or more technology products on pilot projects before making a decision about standardising on one of them for further schemes. If the first pilots do not go well or are inconclusive, then they may try similar pilot exercises with other vendors' products. Alternatively, they may wait for other, similar businesses to test the technologies and identify the benefits first; only then will they select a vendor of their own. Or, if particularly cautious, they may wait for de facto market leaders and standards to emerge before selecting a vendor and product, benefiting from any rationalisation of the market that takes place.[22]

Such conservative customer approaches may yet still have an impact on the UK market. For example, those vendors whose products tended to be rejected in early pilot schemes may find themselves struggling to win new work against those that were successful (on the other hand, if initial tests proved inconclusive, tests with alternative systems may yet show that the original choice was in fact the best one). A strategy of waiting for market leaders to emerge may see the field narrow and an individual provider, or a small number of tried and tested providers, begin to dominate, but this can also pose particular challenges for the vendor(s). As market leaders emerge and customers begin to migrate towards them, the sudden surge in the number of new customers' projects can test the scalability of the vendor's resources, both human and technological. A vendor delivering its software on the ASP model, for instance, may find that its service to all its customers becomes slower or erratic if it cannot anticipate and manage its new workload. Similarly, vendors of traditional customer-hosted solutions may still be able to deliver the software quickly, but their implementation and customer support resources may be stretched beyond breaking point.

Another factor in the market shake-down will be how the different vendors deal with 'churn': customers switching software services. Particularly for ASPs, this is an interesting challenge. Churn manifests itself more easily when customers have not had to invest large sums in the incumbent system and can switch to an alternative supplier with relatively little disruption to their ongoing business; such switches will also be easier if more of the providers are converging on particular industry standards. The phenomenon means ASPs may benefit as customers move from other providers, but it also means their customers can move in the opposite direction.

How the software vendors respond to such challenges will, to a large extent, determine which of them remain active within the AEC market in the United Kingdom and which emerge as the de facto market leaders and standard setters.

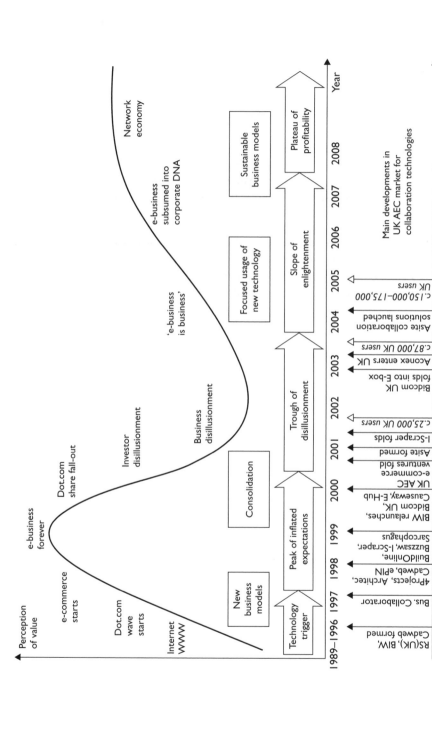

Figure 3.1 The construction collaboration technologies hype cycle (with acknowledgement to the Gartner Group).

3.5   Chapter summary

This chapter has presented a snapshot of the UK market for construction collaboration technologies in 2005 and given some pointers on the key questions customers might use to identify the key differences between the principal players. Given that many commentators still believe there is scope for the number of vendors to reduce over the next two or three years, it has not attempted to provide detailed descriptions of them; instead, it has focused on some of the key generic areas that potential customers should examine as part of their purchasing decision. As the construction market rightly places a high premium on track record, it urges buyers to assess the strengths and UK-specific experience of each vendor's management team and support staff. It also examined some of the key issues relating to the software's origins, and to how it is licensed. And as the technologies will usually be employed to support multi-million pound projects that may last many months if not years, it repeatedly stressed the need to ensure a vendor is financially stable and working to a sustainable business model.

However, it needs to be stressed: a careful review of the different vendors is only part of the buying decision. Later chapters will look at other issues, including the features of the different vendors' systems; in the meantime, Chapter 4 looks at how the technologies can be hosted, and focuses in particular on the pros and cons of the ASP model, widely used in the UK construction collaboration technologies market.

Chapter 4

Hosting construction collaboration technologies

This chapter:

- outlines traditional approaches to software delivery and support, and contrasts this with the ASP model;
- reviews different hosting options;
- looks at the secure infrastructure that might be employed to support an ASPs services.

As outlined in Chapter 3, one of the fundamental differences between some of the main providers of construction collaboration technologies relates to how or where the system is delivered and hosted. On the one hand, there are suppliers who – as ASPs[1] – tend to manage the software and related data on computer systems which they either own themselves or over which they have control (e.g. 4Projects, BIW, Cadweb, Sarcophagus). On the other hand, there are suppliers who prefer to deliver systems where the software and project data are maintained on customers' own computer systems (e.g. Bentley ProjectWise; Union Square, although more focused on enterprise portal systems, offers some project collaboration functionality as a useful extra) or those of their internet service provider (ISP). Some suppliers support both ASP services and customer hosting (e.g. Aconex, Autodesk Buzzsaw, Business Collaborator, Causeway).

This chapter outlines traditional approaches to software delivery and support, and contrasts this with the ASP model. In the context of typical UK construction teams, it then reviews the different hosting options, from hosting in-house by a team member to situations where the ASP, an ISP or a specialist 'managed hosting' provider takes responsibility for managing the delivery infrastructure. It also outlines the advanced infrastructure that an ASP might deploy via a managed hosting provider, and examines related security issues.

4.1 Traditional software delivery

For many years, the vast majority of computer users expected their computer applications and data to be readily available to them locally. If they were working on stand-alone machines, the software and data would be stored on the computer's hard-drive; if working on a corporate network, the application and data might be managed from the organisation's central servers. By and large, software was regarded as a 'shrink-wrapped' or packaged commodity developed by software vendors and sold off-the-shelf to be implemented by the end user or by an in-house IT department.

However, the nature of computer software began to change in the late 1990s as vendors started to deliver applications via the internet. First, vendors started to offer software as a 'downloadable', cutting out the cost of floppy disk or CD manufacture, packaging, distribution and hard-copy documentation – a saving that could, of course, be partly shared with the customer. However, this approach still had drawbacks. For instance, the download process could be time-consuming if the application was large or the internet connection was slow or erratic; the transaction often involved the customer paying a (sometimes quite significant) licence fee before the software could be used (some vendors, of course, offer free evaluation periods after which the user has to pay a fee); there were concerns about online payment and whether individuals had corporate authority to purchase licences; and the application stored on a customer's local machine(s) could soon become out-dated if the customer did not react when a new version, updates or 'patches' became available.

Second, as well as being used on a one-off basis to download software, the internet also increasingly became a vehicle through which users could repeatedly access and use remotely hosted applications and data (e.g. web-based email services such as Microsoft's Hotmail, online banking or travel booking services, online shopping, etc.). To some extent, this reflected the needs of a changing world in which workers were becoming increasingly mobile (so-called 'tech-nomads') and were no longer able or willing to work from just one location. A degree of flexibility could, of course, be achieved by equipping staff with laptop computers, modems and suitable telecommunication links (e.g. virtual private networks (VPNs) allowed mobile workers to penetrate their employers' firewalls and log-on to their networks remotely so that they could download email and synchronise shared data securely). But such infrastructures were often difficult and expensive to set up, administer and maintain, and meant extensive duplication of applications and data on corporate servers and end-users' laptops – a corporate IT department might be required to support literally 1,000s of copies of some applications.

As has been described, a more flexible alternative was to use IP networks where users could interact with software packages and associated data via a standard internet browser. Instead of everything being hosted on an end-user's local machine, the applications and relevant data are hosted remotely and made available to the end-user 'on demand' via the internet. In the UK construction industry context, this alternative also attracted particular interest. As mentioned already, most AEC project teams are temporary, fragmented, multi-disciplinary and multi-company in complexion, and many AEC organisations were often reluctant to allow outside parties to access their corporate systems through their firewalls. Arguably, while such attitudes prevail, effective supply chain networks will have to rely on applications securely hosted outside corporate firewalls. Making the software available via the internet also appealed to many project team members as it freed many individual businesses from having to purchase, download, install and manage additional software applications. However, this new approach can be seen as a threat to the role and responsibilities of conventional IT support staff.

4.2 Traditional IT support

For most business users, then, software packages tended to be purchased off-the-shelf and implemented and supported by either in-house or external IT specialists. The cost, efficiency, knowledge and expertise of in-house IT support may require staff support for multiple systems that have been acquired piecemeal over several years; IT staff often need to be multi-skilled 'jacks of all trades' capable of installing, operating, maintaining and trouble-shooting different hardware and software of different vintages. They might need to be able to support multiple versions of the same software across the organisation. And they might be required to integrate different applications – both standard 'shrink-wrapped' off-the-shelf packages and home-grown programs – on different operating systems or using different programming languages, so that information can be securely shared between different staff, departments and disciplines.

This already formidable task becomes even more difficult and expensive when the IT department must also configure, maintain and ensure the security, availability and reliability of hardware, networks and telecommunications links between different organisation locations, protecting them against extremes of heat, dust, fire, water, theft, physical abuse, etc. (and, as mentioned above, cater for the increasing number of mobile employees who need to work from home, site offices and other locations). To cap it all, particularly as businesses begin to work collaboratively with other members of their supply chains, their IT staff might also be required to implement and maintain secure and reliable links between their systems, those of customers or other business partners, and the internet as a whole.

Moreover, the software applications employed have hitherto tended to focus on supporting internal processes, and, therefore, just the organisation's own staff. As we have seen (Chapter 2), new twenty-first-century business challenges emphasise the need for more efficient and effective relationships with customers and other partners up and down the supply chain. Organisations increasingly need software capable of supporting external users and linking with the systems they use too.

Efficient management of an increasingly complex IT infrastructure can be critical to an organisation's success, but for the vast majority of organisations involved in the AEC sector, it is not usually a core competence. IT, therefore, is an overhead that needs to be carefully managed.

As already mentioned, when looking at construction collaboration technologies, businesses can either follow their traditional route and purchase a software solution that they can add to their in-house portfolio of hosted solutions (and increase the demands made upon their existing IT resources), or they can opt to purchase an 'on demand' service from a vendor functioning as an ASP. Before we assess the pros and cons of the ASP model, it is worth looking briefly at the local hosting option.

4.3 Hosting construction collaboration: the do-it-yourself (DIY) approach

For many construction professionals, particularly those unfamiliar with the nature of web-delivered collaboration software, the idea that all project or programme drawings and documents might be managed by an IT business in a remote data centre can be disconcerting (in the early years, the risks appeared even greater as some ASPs were hosting UK project data overseas – in the United States, for example). Superficially, the idea of retaining the technology and related data locally under the direct control of a team member can seem much more attractive. For a start, local access to data can be very fast, they may feel it will be more secure,[2] and there are fewer legal issues about data ownership.

However, there are some fundamental issues for a client and its project team to discuss if they are to go down this route. For a start, they will need to agree which of them will take responsibility for hosting the software and its related data, and on what terms. For example, the client, particularly if it is a one-off project, may see collaboration as the part of the job of the project team. Designers and contractors may be reluctant to take on the responsibility in case something goes wrong and they end up liable for the system's failure. Some team members may be worried about putting all project data in the control of one team member – particularly if that business got involved in a dispute with other team members, left the project or perhaps ceased trading. There can be issues about copyright of designs, confidentiality and ownership of data. And team members will also need to agree what happens to the data after project hand-over.

Assuming a project team can agree which member will take on the hosting role, the scale of the challenge can also be a deterrent. The tasks involved in planning, designing and constructing a capital asset that may be worth millions of pounds will generate large volumes of drawings and documents, many of which will be business-critical. Within the UK AEC industry, perhaps only a few in-house IT departments will feel they have the technical facilities, resources, skills and experience needed to deliver a service that would meet most clients' minimum requirements. After all, they would be expected to ensure that the whole project team has fast, reliable and secure access to the system at all reasonable times. Could they provide responsive, timely support and guaranteed uptime on all hardware, software and internet connectivity? Could they support their firm's project responsibilities through legally enforceable service level agreements (SLAs) (see Chapter 7 for more on SLAs)? Will their insurance providers provide them with cost-effective cover for this additional responsibility?

These are major challenges that, perhaps, only large businesses can meet. When Autodesk launched its locally hosted Buzzsaw Server Edition in the United States in February 2004, it identified that the product would tend to appeal to larger organisations which 'due to their size, ... have large information technology budgets and IT departments that require more control, security and business continuity. These companies also have significant existing technology investments...' (Autodesk 2004).

4.4   The emergence of 'software as a service'

If a business decides that the challenge of hosting construction collaboration software itself is too great, the alternative may be outsourcing. While many organisations have devolved responsibility for part or all of their IT infrastructures to businesses that specialise in offering IT support, others have been nervous about losing control over key business processes and data. In the UK AEC sector, most businesses continue to manage their IT systems in-house, but (as outlined in Chapter 2) from the late

1990s' forwards, a growing number began to weigh up the pros and cons of using construction collaboration technologies delivered on the ASP model.

The term 'application service provider' became increasingly commonplace, and not just within construction, during the late 1990s. Scott McNeally, Larry Ellison and Bill Gates (respectively CEOs of Sun Microsystems, Oracle and Microsoft) all predicted that businesses would be buying less software and hardware in future; instead, they forecast, we would be renting applications from a software provider and hosting them on servers run by third parties. Microsoft even began suggesting 'software is a service, not a product' – a definition echoed by the UK Institute of Directors (2002) – and it and other major software vendors of conventional, customer-deployed and – managed software (e.g. PeopleSoft, SAP, Siebel) began to suggest customers could build a more direct relationship with the developers of the applications (e.g. enterprise resource planning, customer relationship management, accounting) they wanted to use.

But not all ASPs are the same. Some providers position themselves as ASPs, but their web services can be crudely adapted versions of existing client/server software packages (either their own or licenced from another vendor), perhaps web-enabled using solutions such as Citrix Metaframe or Microsoft Terminal Server. And just as with the phrase 'collaboration', the picture is made no clearer by a lack of agreed terminology.

There is a need to distinguish between packaged software (one-to-one client/server applications) migrated to one-to-many ASP offerings and applications designed from the outset for internet-hosted delivery. Both can be described as ASP, so some industry analysts have tried to coin new phrases to describe the 'net-native' companies. For example, IDC and Summit Strategies have called them business service providers (BSPs) or internet business service providers (IBSPs), while Ovum describe them as 'wave two' ASPs – a description we will use here. Table 4.1 outlines the distinctions between packaged software, migrated client/server applications and BSP offerings.

The 'wave one' ASP model can be viewed as a type of outsourcing: customers get the benefits and reduced risks of offloading complex and expensive IT management to the vendor, but retain control and visibility of their business processes. While the vendor can develop some deployment and management efficiencies, perhaps by spreading day-to-day infrastructure costs across several customers, it is still bound by the performance, security and upgrade constraints of hosting existing solutions originally designed for in-house, one-to-one implementations. This may not always be efficient, particularly if the application has been customised for a customer – in extreme positions, the vendor may have to host one customer system on one server (single-tenant architecture) – whereas a 'wave two' ASP can have dozens of customers hosted on a single server (multi-tenant architecture).

As we saw from the discussion of software licensing in the previous chapter, 'wave one' ASPs also need to adjust their whole business finance models to reflect the less volatile but more predictable flow of revenues, and this can be a difficult transition for a vendor shifting from traditional software sales to ASP software delivery. It will also need to revise how it rewards its sales staff, build some new skills sets within its software development teams to support ASP hosting, and develop a service-oriented culture as opposed to a product-oriented culture.

Table 4.1 Packaged software, migrated client/server applications and BSP offerings

Packaged software	Migrated client/server or 'wave one ASP'	BSP or 'wave two ASP'
Design point		
Designed as product for one-to-one delivery	Retrofits existing vendor product for one-to-many delivery	Built for one-to-many web services delivery
Deployment and management		
Customer	ASP and customer	BSP
Time to deploy		
Months	Weeks	Days
Pricing model		
Customer buys licence, systems, etc.	Customer buys licence, pays subscription fees for hosting, management, etc.	All-in-one subscription fee
Upgrade cycle		
12–36 months; uneven distribution	12–36 months; more controlled distribution	3–6 months; automatic distribution
Customisation		
Unlimited, but costly	Templates to tailor configuration	Self-service configuration wizards
Customer feedback		
Indirect (via systems integrators, resellers); delayed	Direct to ASP, indirect to software developer; delayed	Direct to BSP; immediate

Source: Adapted from summary by Summit Strategies Inc., reproduced in McCabe 2003.

4.5   Benefits of using 'wave two' ASPs

'Wave two' ASPs, by contrast, have built their businesses on a utility or service model, have established their software development, sales and consultancy teams accordingly, and, so long as they can manage their 'churn' rates effectively, can be reasonably confident about their business's financial performance. However, the benefits of ASP delivery extend well beyond financial performance. 'Wave two' ASPs support thousands of end-users on a single common code base, and by focusing on the browser platform they virtually eliminate client-side hassles. As we shall see in section 4.5.1, the model also offers several other major efficiencies.

4.5.1   ASP benefits

Combining the role of developer and supplier benefits the ASP, and they can then translate these gains into benefits for their customers.

- *Faster, more cost-effective development and testing of new software functionality*
  Many conventional applications include hundreds of features that most users rarely if ever use; ASP tools tend to be designed for ease-of-use to meet the very specific requirements of their end users and so lack this 'feature bloat'. Moreover, the application only has to work on a limited range of browsers; it does not have

to work on hundreds of different hardware/software combinations (testing and debugging conventional client/server software can take *twice* as long as it took to write the initial application), meaning that more research and development activity is focused on feature development and enhancement – allowing the software to get better faster.

- *Faster initial deployment of applications* Accessed only via the internet, there are no physical media (e.g. CDs) to manufacture and distribute, nor any installation processes requiring scripts tailored to hundreds of different hardware/software combinations.
- *Little or no customisation of the software required* Customers differ from one another primarily by their data, so most customer-specific requirements can be managed during implementation, normally by configuration of the existing software, to avoid the accusation of them being 'vanilla' applications.
- *Faster and more frequent upgrading or – where necessary – 'patching'* ASPs need only change the code on their application server, and the new feature is instantly available to all users, effectively ending their software obsolescence problems. In client/server scenarios, code would need to be upgraded on dozens, hundreds or even thousands of PCs across each organisation by either in-house IT staff or, if permitted, by the end-users installing CD- or internet-distributed 'patches'. Upgrading an enterprise application might be regarded as successful if it takes less than 100 days; by contrast, even major ASP upgrades can be implemented in under a week.
- *Greater economies of scale* Most ASPs effectively support dozens of customers and thousands of end-users with one code base; they allocate server power over many customers, cutting their hardware and system administration costs.
- *More efficient, expert and cost-effective software support* There are great 'economies of skill'. ASP staff are experienced and familiar with the most recent versions of the application: that is, uniform instances of the ASP's own code that have been implemented on, essentially, a standard infrastructure (note: this familiarity may be compromised where an ASP relies on third-party applications for parts of its solution). Moreover, their expertise will usually be more widely based, drawing on experience of dealing with multiple customer accounts, with costs shared across the customer base.

4.5.2 Customer benefits

Proponents of the ASP approach suggest that it will release customers' IT, training, technical support, and customer service staff from the onerous chores involved in installing, upgrading, and managing traditional in-house applications:

> For instance, a new 100-person company would spend approximately US$200,000 the first year, implementing a group messaging and calendaring system (which would include software installation, maintenance, training, implementation, testing, consulting, updates, and other things), but the same system outsourced could range from US$30,000 to US$55,000 in annual costs. This would allow the company to apply cost savings to its core business and free up IT personnel.
>
> (Wood 2001)

In more detail, the benefits to the customer can include:

- *Lower cost software* The ASP model was envisaged as bringing costly enterprise-class applications within reach of small/medium-sized enterprises (SMEs) and its lower cost approach is also attractive to larger businesses.[3] Many of the factors covered in Section 4.5.1 combine to significantly reduce the cost of delivering applications on ASP platforms compared to client–server systems. The saving can be substantial. In the construction collaboration technology market, for example, US-based Constructware estimated that it costs users about 60–80 per cent more to get the same features on a client–server platform as it does in an ASP environment (Unger 2001).[4]
- *More cost-effective provision implementation, training and ongoing support* Since it seeks a long-term relationship, the ASP is incentivised to maintain a good quality service (in a pessimistic view, the conventional software vendor, by contrast, having got an up-front payment, almost wants to ensure the customers are *not* happy, so that they can be sold an upgrade in, say, two years time).
- *Little/no capital outlay* With hosting looked after by the ASP, the main issue is ensuring access. As this is normally via a standard internet browser, few organisations will need to invest in new hardware/software to access ASP software.
- *More predictable expenditure on software, hosting and support* As discussed in Chapter 3, the ASP model tends to deliver applications on a service model paid for by regular monthly subscription payments.
- *No depreciation.*
- *More flexible* The service can be turned on and off quickly.
- *Faster initial deployment of applications* Less end-user and IT staff time is spent on installing and debugging software on individual computers; this may also extend to less reliance on systems integrators and consultants.
- *More direct and responsive approach to new or evolving customer requirements.*
- *Fewer supplier excuses*

  > The pure, native ASP can never, ever point the finger at someone else. Bugs are its fault. Service interruptions are its fault. There is nowhere to hide. Sounds like a disadvantage? It's not. A company that is forced to listen to its customers and solve their problems will deliver superior services and have happier customers in the long run.
  >
  > (Christian 2003)

- *Faster development and release of new software functionality, upgrades and/or 'patches'* An ASP can deliver frequent and more 'digestible' upgrades over 3–6 months, while conventional software vendors tend to produce more infrequent, major upgrades ever 18–24 months.
- *Reduced risk* Few organisations can commit to 24×7, fully resilient and secure management of their IT infrastructure;[5] an ASPs core activity is 24×7 provision of the application and associated data to its customers.
- *Greater availability of the application* Unlike most in-house departments, an ASP will usually have an SLA guaranteeing high levels of application availability and reliability; if the service is interrupted, the ASP would probably be much quicker at reinstating normal service than an in-house team.

- *More scaleable* The ASP, not the customer, is responsible for ensuring adequate hardware and bandwidth to cope with current and anticipated levels of application usage.
- *Better use of in-house IT resources* Application-specific IT skills can be out-sourced to the ASP's own experts, leaving the in-house team to focus on its other responsibilities. The ASP's one-to-many, multi-tenant system architecture also means new features can be quickly and easily deployed without touching any net-works, servers or PCs, reducing the customer's total cost of ownership (TCO) when compared to a client–server application.
- *Better staff productivity and working environment* In addition to the more focused work of the IT team, individual users in departmental or functional roles will enjoy greater access to the information they need – external as well as internal – to do their work efficiently; being accessed via a web browser means the application and data can also be accessed from anywhere, supporting more mobile working. IT inefficiency will be less of a barrier to recruiting high-calibre people to the organisation and the industry within which it works.
- *Greater protection against viruses* ASPs usually have dedicated technical resources devoted to maintaining state-of-the-art anti-virus systems, while the web-based platform itself also prevents viruses being distributed (e.g. many viruses exploit email systems, but most construction collaboration technologies do not directly receive and process email, while uploaded files are never automatically executed – a common release mechanism of computer viruses – on an ASP's user-facing 'production' servers).

4.6 Traditional software or ASP: other considerations

The benefits of the ASP approach may appear very persuasive, but any organisation considering a relationship with an ASP should examine its own IT situation carefully and, in particular, it should weigh up its approach to risk management. There are also wider technological trends to consider.

At a strategic level, for example, a business might need to look at:

- *Sunk costs* The investment it has already made in its people and existing hard-ware and software.[6] For example, some businesses may have invested in develop-ing their own solution or perhaps adapted a client–server package so that it can be accessed via the web (e.g. via Citrix or Microsoft Terminal Server solutions), adding another layer of servers and software to the IT infrastructure, plus licensing and support costs.
- *Risk* As already mentioned above, any internal issues it may have regarding control, security and availability of the application and data (sometimes resolved by a review of the ASP's hosting arrangements – see Section 4.8.1). The timing of an internal system upgrade, for example, can be easily controlled, but an upgrade to an externally hosted, multi-tenant system may not occur at the optimum time for all its end-users – again, it pays to research the frequency and duration of upgrades or maintenance breaks, and to make sure these are covered in any SLAs.

- *ASP financial stability* Perhaps most important, it will need to be satisfied about the strength, experience, financial security and independence of its proposed provider (covered in Chapter 3) and of any third party the ASP uses to support its service (see Section 4.10).

At a more practical level, meanwhile, the organisation will also need to assess the likely impact on its existing IT resources, whether in-house or outsourced. For example:

- *Impact on IT staff workload* Existing staff will probably be the first port of call to manage web browser settings, network configurations, and connectivity and access issues relating to the organisation's ISP (see also Chapter 6) – issues that will normally be resolved in conjunction with staff from the ASP. Switching from supporting an in-house client/server application to an externally hosted ASP solution will, however, usually mean that some staff can be redeployed to more strategic IT issues.
- *Integration with legacy systems* ASPs vary in their ability to integrate their application services with a customer's existing systems; buyers will need to clarify the cost and complexity of any required integration work.

In some respects, the emergence of the ASP model came at a bad time – just as the dot.com bubble reached its height. When the bubble burst, dozens of over-ambitious and over-stretched ASPs joined the rest of the dot.com sector in financial free-fall. Michael Christian (2003) summed it up:

> The pure ASP builds its own applications solely to provide services. This creature lives in a very tough ecological niche. It is by definition a start-up. It must build, host, deliver, manage and support its own applications. During the technology boom from 1995 to 2000, many were spawned. Few survived.

However, the legacy of the dot.com crash has been a hard core of well-run ASP survivors.[7] At the same time, declining IT budgets have forced many potential software buyers to review their IT expenditure. Hard-pressed IT directors wanted to squeeze rolling licence fees and reduce up-front implementation costs, forcing some traditional software vendors to offer new ASP routes to market for their existing products (though such routes will typically be as inefficient, 'wave one' ASPs – see Section 4.4).

4.7   Minimum quality of service (QoS) considerations

After weighing up the pros and cons of the ASP approach versus traditional software, the customer should be able to decide which option to adopt. If he or she has decided against self-hosting and has selected the ASP option, the next step is to assess the infrastructure through which the ASP will deliver the service.

ASPs differ in how they support their services but it is useful to agree some basic QoS requirements; for example:

- *High levels of performance* For example, speed of access to data.
- *Constant availability of application and data* 24 hours a day, 365 days a year, coupled with low latency (i.e. minimal delay between someone sending data and the recipient receiving it).

- *High reliability and resilience* For example, full redundancy (all primary network system components such as servers, power sources and telecommunication links have secondary back-ups in case they fail) and detailed disaster recovery plans in case of a catastrophe (e.g. a provider may undertake to replicate its service, complete with all data, at another hosting location within, say, 72 hours).
- *High levels of security* Both physical (preventing unauthorised access to server racks or other equipment, and managing risks associated with fire, flood, explosion, etc.) and technological (preventing hackers, viruses, distributed denial of service (DDoS) attacks, etc.).
- *High levels of service management* Ideally, recognised standards of best practice (e.g. the UK government-created IT Infrastructure Library (ITIL), increasingly adopted as the standard for best practice in IT service provision).
- *High levels of scalability* Customers need to be sure that the ASP is anticipating future hosting capacity requirements (maintaining adequate service levels irrespective of the loads imposed by growing numbers of projects, users and logins, increasing volumes of information stored, etc.).

4.8 The external hosting options

With these minimum requirements in mind, would-be customers can begin to assess the merits of each ASP's infrastructure. How each ASP meets these requirements will depend on the resources they have available and the importance they attach to each one. The options facing UK clients and AEC organisations will usually reflect the locations of the servers used to manage the applications and their underlying data. For example, server(s) might be:

- managed in the ASP's own facility;
- located in data centre run by an ISP – options include: sharing a server with other (sometimes competing) applications and/or websites; a dedicated server leased from the ISP; and co-location – the provider places its own server, set up with all the software, in the ISP's facility;
- managed at a specialist hosting/data centre ('managed hosting' – see Section 4.8.3).

Moreover, any of these options might have a further geographical dimension. Particularly in the early days of UK construction collaboration technology, scare stories circulated about ASPs who hosted UK project data in facilities based in the United States or other countries. This perhaps reflected poor bandwidth performance, fears about different data protection regimes or worries about the additional costs of dealing with different legal frameworks should an ASP or its host go bust.

4.8.1 In-house hosting by ASP

Hosting by the ASP might seem the obvious alternative as it will presumably be developing its solution based on detailed knowledge of its own hosting environment's capabilities (though there is, of course, a danger that software development might be guided more by hosting capability than by customer demand or market forces).

Where an ASP hosts its own technology in-house, customers need to be sure that the ASP has the technical facilities, resources, skills and experience needed to meet their minimum QoS requirements. Just as the core activity of most AEC businesses is not IT delivery and support, the forte of most ASPs is software development, not creating, maintaining and, where necessary, scaling-up a secure, reliable and robust 24/7 data facility. As touched on at the end of Chapter 3, should the business suddenly experience a surge in demand for its services, customers need to be sure that the provider has the human and technical resources needed to meet the increased demands made.

4.8.2   Hosting by an ISP

If the system is hosted at a conventional ISP, customers should be aware that some ISPs may not provide pure IP networks. Their communications links may segment into other services (e.g. voice or fax), so that data shares bandwidth with non-IP traffic, compromising quality and reliability. Also, ISP networks hosting public dial-up operations can be subject to load fluctuations; by contrast, networks providing hosted applications with dedicated bandwidth 24/7 should be capable of routing very high volumes of traffic at consistently high data rates.

Also, many ISPs concentrate on selling network bandwidth capacity on their global networks, and providing power and rack space in their data centres. As a result, they often manage a mixture of hardware systems and operating systems, juggling bandwidth and compromising as they try to reconcile the competing needs of website owners, service providers, etc. Moreover, however sophisticated the data centre, hardware and software, it still needs careful, expert management by certified professional staff 24/7 – people who, ideally, specialise in a small number of hardware platforms and operating systems. After all, it is these people that must devise, implement and adhere to the procedures that underpin the integrity and security of the system.

Finally, a concentration on the hardware, power, connectivity and infrastructure management issues of hosting also obscures the most volatile part of the whole hosting solution: the web application itself. Efficient hosting of collaboration technologies requires proactive anticipation, monitoring and management of the ongoing demands made on that infrastructure so that it can be expanded as the volume of traffic and content grows.

4.8.3   Managed hosting

Clearly, then, there can be potentially serious drawbacks with services which rely on self-hosting, or hosting on an ASP's own servers, or on use of a conventional ISP. Hosting may require non-core or complementary skills, expertise and IT resources that are not available within the customer organisation, the ASP or a traditional ISP. Recognising this, several UK construction collaboration providers have opted for 'managed hosting' (e.g. 4projects used Leighton publishing group sister company The Data Corporation then BT, BIW uses Attenda, Causeway uses Globix).

Within a dedicated, purpose-built data centre, a managed host procures, configures, installs and maintains the necessary servers, firewalls and other devices that its customer (i.e. the ASP) requires for its software architecture. Once configured, the host then provides dedicated bandwidth for the ASP's applications, and connects the servers to the web via its own network, where the application is constantly monitored to ensure availability and optimal performance. Normally, the ASP does not get involved in the maintenance of the hardware, but is able to take advantage of the host's processes and their enshrined expertise and experience to implement, maintain and update services in less time and with higher reliability than other hosting methods. Moreover, managed hosting tends to ensure that technology services remain easily and seamlessly scalable. This is vital as the user base grows, as demand for bandwidth increases, as the applications are upgraded, etc.

4.9   A hosting data centre

When looking at the details of hosting arrangements, therefore, customers should assess the ASP and its infrastructure (and any third party provider) against the minimum QoS requirements. Some ASPs will invite potential customers to visit the data centres where their applications and associated project data are hosted; such visits can give first-hand experience of the technologies and people deployed to ensure that the ASP's service is as secure, reliable and robust as possible.

A data centre will typically display most, if not all, of the following features:

- *A secure, controllable physical environment*

    i   Extensive environmental controls with mechanical rooms situated separately from the server areas.
    ii Key-cards, electromechanical locks and biometric palm readers to ensure that only those authorised to enter, and then authenticated, can gain access.
    iii Closed circuit television surveillance, augmented by security personnel, to constrain and monitor movement within and around the facility.
    iv Continuous monitoring by fire suppression systems for smoke, chemicals and other hazardous materials.
    v   Protection from physical terrorist attacks or other catastrophes (blast-proof facilities, etc.).

- *Fully redundant services (telecommunications, power, etc.)*

    i   Multiple partnerships with telecommunications businesses (telcos) to provide pure IP-routed access to internet backbone, plus secondary backbone connections.
    ii Peering arrangements with each telco to ensure data centre traffic is prioritised (peering is the process of linking one telco's backbone network to another's to allow traffic to travel across the networks), helping maintain guaranteed, dedicated IP bandwidth.
    iii Primary and backup power supply systems, include redundant uninterruptible power supply (UPS) systems as well as multiple secondary generators.

- *Robust IT hardware*

    i   Primary and back-up firewall(s) and other security hardware and applications.
    ii Router(s) and other communication equipment connecting servers to the internet, plus back-up connections.

   iii   Racks, cabling and server-friendly facilities.
   iv   Primary and secondary backup servers, etc.

- *Expert, pro-active personnel*

   i   High-calibre infrastructure managers, working to accredited information and IT management standards (e.g. BS7799/ISO17799 and ITIL).
   ii   Specialist in-depth expertise on specified hardware and software platforms.
   iii   Remote application monitoring and support.
   iv   Capacity management and planning.
   v   Dedicated anti-virus resources (e.g. real-time monitoring of files uploaded to system, immediate implementation of anti-virus updates, etc.).
   vi   Change management.
   vii   Management of software upgrades and patches.
   viii   Database back-up, maintenance, off-site storage and recovery processes.

This last item is particularly important. The ASP and/or the hosting provider may experience a catastrophic failure during which some or all of the project team's data could appear to be lost. To prevent such loss, a rigorous system of regular back-up and retention must be employed. For example, a hosting provider may take a 'snapshot' of all the project data on a daily basis and store two copies of the tapes for 30 days in a fireproof safe located in a secure environment away from the data centre. In the event of any loss of data, all client data up until the previous backup will be available for immediate restoration.

4.10   Investing in security

It is dangerous to underestimate security issues.

> Information is the lifeblood of today's business, underpinning day-to-day operations and facilitating effective decision-making. Increasingly, access to the right information by the right people is vital to gaining competitive advantage or simply remaining in business. To provide this access, businesses need to understand the associated risks and put in place appropriate counter-measures.
>
> (PriceWaterhouseCoopers/DTI 2002)

A 2001 Communications Management Association survey of IT professionals indicated that one in three UK businesses had been the victim of a major security break-in; the 2002 PWC/DTI survey put the figure higher, at 44 per cent. Almost half said that the future of their organisation could be ruined by a serious hacker attack. Of 167 participants in a one-month BBC experiment in 2001, 159 were subject to hack attempts. Over 1,700 hack attempts were made, with one participant subject to 91 separate attacks. Managed hosting provider Attenda constantly monitors intrusion attempts, and its experience suggests that a network will typically face approximately 35,000 virus attacks or attempted hacks *every day*, many from automatic probes by software robots programmed to ceaselessly search the internet for vulnerable machines. This is borne out by experiments such as that run by *USA Today* and Avantgarde in which six unguarded computers suffered 305,955 online attacks in two weeks, with one computer penetrated and compromised within four minutes of going online.[8]

Full protection against virus attack is critical. Ideally, the hosting provider will have a comprehensive virus protection system, with virus detection files updated regularly (e.g. nightly). Such a regime can ensure that providers are protected against new virus threats almost before they become public knowledge. Similarly, a hosting provider will also need to ensure that its firewalls provide adequate protection to the ASP's service infrastructure. This should extend to monitoring for unusual traffic patterns indicating 'denial of service attacks' or hacking attempts which can then be blocked.

A small minority of the UK construction collaboration providers have systems that carry the BS7799 (ISO 17799) accreditation for internet information security. BS7799 requires, among other things, that businesses implement change control and business continuity plans, making websites more robust and protecting them from potential threats to availability. Companies need to either have an in-house team dedicated to making this practice a reality, or work with a partner that can provide the necessary expertise and time to pre-empt disasters before they happen. (Cadweb has established an in-house BS7799 capability, while BIW's managed hosting partner, Attenda, is accredited.)

However, even the most robust protection against intruders, viruses and denial of service attacks will not offer complete security. Access to most construction collaboration systems is normally reliant upon an end-user entering a valid password, but, clearly, if the password is given away then the system is not protected. As PriceWaterhouseCoopers/DTI (2002) identified: 'People are often the weakest link for security. …many UK businesses are spending considerable time, effort on money on implementing sophisticated technology, without developing a security awareness culture within their organisation to support it.' The ASP's client and other members of the project team therefore have a responsibility to ensure internal procedures are in place to secure the confidentiality of these logins to guard against unauthorised access.

4.11   Financial security of hosting providers

As well as understanding the security of an ASP's infrastructure, it is vital to establish that any third party company providing the infrastructure is itself stable and secure.

During the early boom years of the internet, many companies – including several major telecommunications businesses – invested heavily to join the market and build new data centres to handle the anticipated demand for ASP and website hosting capacity. But when the 'dot.com bubble' burst in 2000, market forecasts proved to be over-optimistic, and the building binge had, in many regions, created a glut of under-used data centre space. Simultaneously, customers became more discerning, shifting away from co-location towards managed hosting. As a result, co-location data centre space became a commodity, with pressure to keep prices low in order to compete in a buyer's market. Sales also fell at a time when it was even more expensive to transform a co-location ISP into a managed hosting provider.

In an overcrowded market, many data centre owners, including some telcos, experienced significant problems. For example, in September 2001, Exodus Communications (an American corporation whose UK facilities hosted some UK collaboration vendors' systems) applied for protection under Chapter 11 of the

US Bankruptcy Code. By the end of February 2002, a substantial portion of its business and assets, including those in the United Kingdom, was sold to Cable & Wireless plc. Similarly, ISPs such as PSINet also faced financial troubles, with PSINet also applying for Chapter 11 protection in early 2001.

4.12   Chapter summary

This chapter has explained the differences between conventional approaches to software delivery and the ASP model. It has also reviewed different hosting options, from hosting in-house to specialist 'managed hosting'. Read in conjunction with Chapter 3, this chapter will help the reader make some informed decisions about the strengths of the different technology providers, their business and software licensing models, and, if they are ASPs, their hosting infrastructures. Chapter 5 shifts the focus towards the applications themselves.

Features and functionality of construction collaboration technology

---

This chapter:

- gives a 'snapshot' of the main features, plus a few optional extras, shared by the leading construction collaboration technologies available in the UK AEC market;
- explains two key distinctions between the AEC industry and other sectors (construction's extensive use of graphical information, and the need to share this among an often temporary and fragmented project team);
- discusses how these challenges have been addressed to date.

So far (Chapters 3 and 4), the main focus of this book has been on helping the reader understand the key differences between the different UK-based technology providers, their business models and their hosting arrangements. When differentiating between the different vendors, track record is often the key factor, so the history, experience and financial status of the vendor will be important, the existence of a substantial UK-based user community and a good list of reputable UK clients can also help narrow the field, and a realistic low-risk approach to hosting is also desirable. At the same time, each vendor's construction collaboration application will need to be able to meet the practical information and communication needs of most UK construction project teams.

Essentially, they need to support the requirements of a multi-disciplinary team, drawn from multiple companies, all based in different locations with their own IT systems, brought together – usually temporarily – to plan, design, construct and, in some cases, to manage the operation and maintenance of an asset or group of assets. They replace localised sets of data held by individual team members or companies with a centralised repository or data store that can be accessed by all authorised team members, usually using a lowest common denominator technology: a computer equipped with an internet browser and a telecommunications link to the internet.[1] To reiterate the point made in Chapter 1, the 'construction collaboration technology' employed is actually a combination of several technologies: a computer, a modem, an internet connection, a web browser and one or more other software applications (i.e. the collaborative software and, in most cases as we shall see, viewer tools used to allow the user to see and, if necessary, interact with the information held in the repository). Assuming the end-user has all the necessary hardware, internet access and a web browser, the focus shifts to the capabilities of the software applications offered.

In addition to capturing key background information about the vendors, some UK trade bodies (e.g. RICS;[2] CICA 2003) have undertaken detailed questionnaire surveys to assess the range of features delivered by the leading providers (similar question-naires have also been used by potential customers as part of their information-gathering or short-listing processes[3]). As most providers will be reluctant to say their system does not possess a particular capability, most of boxes get ticked – even though, in some instances, the capability may only be delivered by a circuitous 'work-around'. The outputs from this type of survey then tend to be presented as a matrix summarising the vendors' responses against each of the questions asked, and there is often little to choose between the different systems in terms of the presence or absence of particular features (see sample page in Figure 5.1). Given the speed of IT development and the competitive nature of the marketplace, such paper-based feature matrices can also quickly become out-of-date, and the value of such paper-based 'snapshot' surveys is limited, as the CICA admitted in its document.

To overcome the qualitative shortcomings of a feature survey, most customers and/or their fellow project team members will seek detailed references from existing customers and end-users, will want system demonstrations so that they can see how the different applications manage particular documents and processes, and will ask for access to demonstration environments where they can try out each application's functionality for themselves, and learn how easy (or not) it is to use each system.

This chapter presents a snapshot of the main generic features of the leading construction collaboration technologies available to users in the UK AEC market, attempting to identify the core areas of functionality that are unlikely to change significantly over time. It also explains two key distinctions between the AEC industry and other sectors – namely, construction's extensive use of graphical information, combined with its need to share that graphical information among an often short-lived and geographically dispersed project team – and discusses in detail how these challenges have been addressed to date. From this discussion, readers should begin to understand the importance of drawing viewer and mark-up technologies within construction collaboration. The chapter also briefly describes some of the areas of functionality offered as options or which differentiate the main systems on offer (as with many new areas of IT, the exact terminology may vary from vendor to vendor, and the range of features may increase over time).

Few of the vendors market their systems on the basis that they are 'out-of-the-box' solutions that can be purchased, installed and operational within minutes. Ease of use is a critical dimension affecting how quickly end-users will accept and become profi-cient at using a particular system, but good software design will be only part of the solution. Providers will usually provide consultancy and implementation services to help the customer and/or the project team to:

- review its needs; then
- compile a project protocol document detailing the pragmatic processes and standards to be applied by the team;

System features

	Sarcophagus	Epin	C3	BC	Bidcom	Cadweb	BIW	4Projects	Architec	Autodesk	Atkins	Causeway	BuildOnline
No. of UK user firms	1,727	500+	Approx 150	60	70 direct paying clients	600	2,500		20+	600+	~250	200	>1,000
No. of active projects	391	60	1	500+	200+	35	1,035	2,500	30+	200+	250+	55	>300
No. of these projects on site	307	50	1	500+	60	30	c500	Not aware		100+	~25	52	>150
No. of user firms worldwide	1,959	500+	Approx 150	N/A	200 direct paying clients	650	2,574	3,000	50+	20,000	~300	245	>3,000
No. of UK user firms who have an equity share or other financial interest in the product	0 (Directors are owners)	0	0		3	0	0	0	0	0	1	0	1

Does the system offer

	Sarcophagus	Epin	C3	BC	Bidcom	Cadweb	BIW	4Projects	Architec	Autodesk	Atkins	Causeway	BuildOnline
(a) Processor programme management	Yes	Yes	Yes	Yes	Yes	Yes	Yes	Yes	Yes	Yes	Yes	Yes	Yes
(b) Document management	Yes	Yes	Yes	Yes	Yes	Yes	Yes	Yes	Yes	Yes	Yes	Yes	Yes

Project set up

	Sarcophagus	Epin	C3	BC	Bidcom	Cadweb	BIW	4Projects	Architec	Autodesk	Atkins	Causeway	BuildOnline
Has the users' project manager total control over the project set up procedure to meet their specific needs?	Yes	Yes	Yes	Yes	Yes	Yes	Yes	Yes	Yes	Yes	Yes	Yes	Yes
Can the structure and settings be closed to a new project?	Yes	Yes	Yes	Yes	Yes	Yes	Yes	Yes	Yes	Yes	Yes	Yes	Yes

Figure 5.1 Sample page from CICA survey (2003) (by kind permission of CICA).

- configure the software features and functionality accordingly; then
- train key individual team members to use these tools;
- provide ongoing support including, say, email and telephone help-desk services.

As the final sections of this chapter suggests, the ease of use of the application and the scope and quality of the vendors' implementation, training and ongoing support services should, therefore, be carefully assessed as these will have a critical bearing on the roll-out of the application to the project team. Once the initial implementation work has been undertaken, end-users should be able to use the application's features efficiently and effectively. Briefly, and for the purposes of structuring this chapter, these features make it possible to:

- *organise* companies, teams and individuals with specific roles, responsibilities and access rights, and to *organise* stored information into types and categories, with relevant statuses, security clearances, etc.;
- *communicate* information to other team members in forms that can be easily accessed and reviewed and that allow easy feedback; in short: to collaborate;
- *manage* and track key tasks and business processes, and their related information, and to build up a complete project knowledge audit trail.

5.1   Organisation features

Before any project team can begin working on a scheme, it needs to be organised so that different companies and individuals know their respective status, roles and responsibilities, and work to a consistent set of rules and processes. The same applies to organising an online team, but with the added challenge of incorporating those roles, responsibilities, rules and procedures into the practical, day-to-day functionality of a software application. The following subsections outline some of the typical features available.

5.1.1   Security settings

- *HTTPS* At a high level, it may be decided that the project or programme warrants especially secure hosting: for example, access using a secure server certificate (HTTPS), commonly used by banks and other organisations to safeguard financial transactions via the internet.
- *Secure log-ins* Typically, the project team will agree what type and level of access will be required by each team member (these and many of other items outlined in this chapter will be usually be documented in the project protocol document). Most construction collaboration technologies require users to log-in before they can access the collaboration platform, so the next layer of security limits access to authorised individuals in possession of a valid username and password (Figure 5.2).[4] As an additional precaution, some systems will prompt users to change their passwords at regular intervals.
- *User profiling* As end-users log-in, their details will be matched to user profile details stored against their names in the application's underlying database, and the data that they see will be configured to suit their access rights, duties, roles

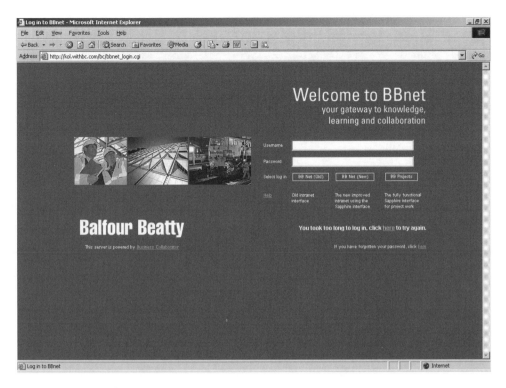

Figure 5.2 Example of customised user log-in page – from Business Collaborator (image copyright to and by kind permission of Business Collaborator Ltd).

and responsibilities. This is one of the most critical areas to be discussed in agreeing project protocols. At a high level, this may mean that some 'super users' gain access to all projects within a particular programme of works, while others are granted access only to one or two individual schemes. Teams will therefore need to agree who these administrators should be, what their responsibilities are, how any confidential information might be managed, etc. At a lower level within a project, the profile settings may mean some documents or processes are not accessible to particular individuals, perhaps for reasons of confidentiality or simply because they concern matters outside those individuals' remit. Not displaying items to which users have no access rights helps to avoid constant reminders of their exclusion.

5.1.2 User administration

* *User directory or address book* All team members should be listed in an online directory, allowing other authorised users to find names, job titles, company details, addresses, telephone numbers, email addresses, etc. (see Sarcophagus and Causeway examples in Figures 5.3 and 5.4). Such directories tend to grow as

Name ▲	Job Title	Organisation	Last Logon	Admin	Logs	Edit	Rem
Administrator, Organisation	test account	Sarcophagus Ltd	20/07/2004	Org.			X!
Administrator, Project	test account	Sarcophagus Ltd	11/11/2004	-			X!
Davis, Richard		GJH Architects	Never	-			X!
Hayward, Jeff		GJH Architects	Never	Org.			X!
Howarth, Graham	Managing Director	Sarcophagus Ltd	17/11/2004	Proj.			X!
Howarth, Scott		Milford Business Consultancy	Never	-			X!
Hudson, George		GJH Architects	Never	-			X!
Lishman, David		GJH Architects	Never	Org.			X!
Loomes, Matthew	Network Administrator	Sarcophagus Ltd	23/11/2004	-			X!
Machan, Philip		Milford Business Consultancy	Never	-			X!
Sainter, Jeremy	Operations Director	Sarcophagus Ltd	24/11/2004	Proj.			X!
Smith, Mark		Milford Business Consultancy	Never	Org.			X!
Stephenson, Richard	Director	Milford Business Consultancy	Never	-			X!
Ulyett, Phil	Developer	Sarcophagus Ltd	24/11/2004	Proj.			X!
Whittaker, Chris	The-Project Support	Sarcophagus Ltd	24/11/2004	Proj.			X!

To add a new member to this project:

1. Select the organisation from the organisations address book
2. If the organisation does not exist you must create a new one
3. Select an existing organisation member by clicking the 'add to project' link
4. If the member does not exist you must create a new one. They will be automatically added to the current project

Click here to go to the organisations address book

Figure 5.3 Example of project team user directory (1) – from Sarcophagus's the-project.co.uk (image copyright to and by kind permission of Sarcophagus Ltd).

a project progresses, with sub-contractors, sub-consultants, suppliers, etc. being added to the original core team; some users (e.g. building control officers) might only be granted short-term access rights to cover their inputs. Some systems may allow the directory to synchronise with other offline applications such as Microsoft Outlook or Lotus Notes.

- *Issue or notification lists* In most teams, individual users can be grouped by discipline, role, responsibility, authority, seniority, etc. The collaboration platforms typically provide for the creation of standard lists that can be used to speed up the distribution of information to a relevant group or groups of users. These will be capable of amendment, allowing users to add or remove individuals from the lists, depending on their needs; they may be 'cloned' to start new lists; and new lists may be created at later stages (e.g. as sub-contractors are appointed and need to be 'kept in the loop').
- *External notification* While some users may log-in daily, others may be only occasional users, and may not be aware that something needs their attention.

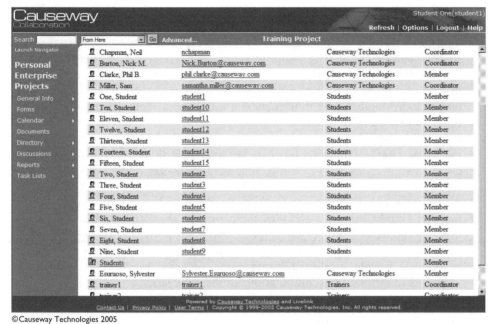

©Causeway Technologies 2005

Figure 5.4 Example of project team user directory (2) – from Causeway Collaboration (image copyright to and by kind permission of Causeway Technologies Ltd).

Most of the systems provide for such individuals to be alerted, usually by email (though fax, text messages to mobile telephones and IM might also be used), prompting them to log-in and undertake the actions required of them.

5.1.3 Information administration

- *Basic project details* The core details of the project (e.g. client name, project name, site location, etc.) will be accessible to all users, along with associated information (e.g. maps, site access plans, etc.). In most cases, customers will also have the option of customising the project log-in page and/or the user interface so that it includes their logo and other elements of their branding (see Figure 5.2 for an example of a branded log-in page).
- *Information structure* Faced with managing a large number of drawings, documents, photographs and other items of information, teams generally prefer to divide this into logical categories. The construction collaboration technologies tend to represent this by using the familiar folder-based structure used to store and manage files on a PC (in some systems storage is actually based on folders, others employ a relational database but retain a user interface resembling Windows Explorer to present categories or 'registers' of information in just the same way).[5] The precise structure and folder/register naming conventions are usually agreed with the team during development of the project protocol document and system set-up; some systems allow for new folders and/or

sub-folders to be defined and created when required, though this right may be limited to certain authorised individuals.

- *Information security* Of course, some files may hold sensitive or confidential information, and it will be necessary to limit access and viewing rights to them. In some systems, these files are visible but cannot be accessed; teams may be wary of systems that allow an unrestricted, global view of all the folders/registers and files in case end-users start exploring; they may prefer instead that sensitive files (either individually, or within particular categories) remain hidden to all but those authorised to see them.

- *Information search* As the mass of drawings and other documents grows, finding particular items of information can become difficult (readers may well have experience of trying to locate files that may have been mis-filed in the wrong folder, or finding sub-folders or files with identical or similar names and content). Accordingly, collaboration platforms normally provide powerful search tools. It may not even be necessary to know the file name; the search tools usually allow users to initiate detailed searches, for example, for particular file types, for files published by particular end-users or companies (or issued to certain users and/or companies), for files published during certain time-frames, or with certain words in their descriptions, plus combinations of these criteria.

5.2   Communication features

As has been mentioned already, traditional construction projects generated huge volumes of paper drawings and other documents, and the advent of computers initially did little to change that. However, as organisations have become more adept at using electronic tools across networks and then the internet, the processes of issuing, viewing, commenting upon, and marking-up have also been transformed. Today's construction collaboration technologies now allow professionals to undertake many of these tasks online. The following subsections outline some of the typical features available.

5.2.1   File publication

- *File issue* This term is used to cover the registration of a drawing or other document file in the collaboration system, that is, the entering of key details – 'metadata' – about the file into the system in advance of its upload (see later). In some instances, details may already be recorded by the originating organisation, either manually or electronically, in an internal drawing or document management system.[6] When publishing files to a collaboration system, the details typically include: the filename (usually entered automatically when the user browses the relevant folder on his or her local hard drive), the title (if different to the filename), a document number and version number or letter (as part of their project protocols, some teams agree document or drawing numbering and version conventions to be used throughout a project; these may be managed automatically be the collaboration system), the document status, the folder/register to which the file is assigned (assuming this is not automatically set), and the names of the end-users to whom the file is being issued. Depending on the system, further options may be available; for example, the file may be associated with a particular zone of the project or relate to a particular work package – such attributes might also be entered at this stage.

- *File uploading* Once all the issue details about the file have been entered into the collaboration system, the end-user will usually click a button on the screen that will initiate the publication process. For some file types – mainly drawings – it may be necessary to upload the file in more than one format (see Section 5.4), depending on the project protocol agreed for drawing formats. For example, an AutoCAD drawing may be published in its native DWG format, in a 'light' design web format (DWF) format, and/or maybe as a plotfile (a file used to drive printers to produce a true representation of a document or drawing, for example, a Hewlett-Packard Graphics Language (HPGL) file). All three files might then also need to be packed into a ZIP file, mainly to compress the DWG, to speed up the process of upload to the collaboration system's servers. As this process of converting native CAD files into DWF, portable document format (PDF), etc., can be quite time-consuming, some vendors provide tools to automate and speed up the task (BIW, for instance, offers designers an AutoCAD plug-in that produces the required files and prepares them for publication to its collaboration system).[7]

- *Time and date-stamping* When files are published to a collaboration system, the time and date, the name of the publisher and his or her company details, etc., are automatically recorded and form part of the audit trail for each file (Figure 5.5).

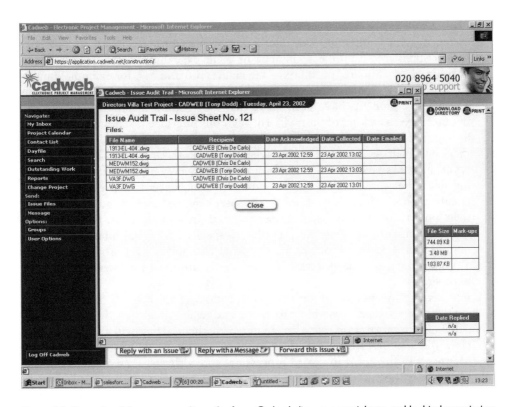

Figure 5.5 Example of document audit trail – from Cadweb (image copyright to and by kind permission of Cadweb Ltd).

- *Batch processes* In many instances, end-users will not be publishing just one file, but several. Most construction collaboration platforms provide batch publishing functions that allow multiple files to be selected for upload, with file issue details entered just once. The batch process may be initiated immediately or it may be timed to take place at a later time, perhaps when network traffic and/or the internet is less busy.
- *Issue notification* Once the publication process has been initiated and the uploaded details and associated files have been received by the collaboration system, recipients who are logged-in will get an on-screen notification that they have been issued with new drawings or other documents (users who are not logged-in may receive an external alert – see 'external notification' earlier – that something requires their attention).
- *Integration with drawing issue management systems, fax, email, etc.* Some systems offer integration with in-house systems (e.g. some architects or engineers use drawing management systems or submittal logs, or log change orders) so that publication can be recorded in both systems simultaneously. There may also be facilities to send and receive faxes, to send emails to non-project participants (and record incoming emails), and even to provide details of hard-copy reference information stored elsewhere.

5.2.2   File management

- *Project summary or home page* Most users normally log-in to their collaboration system at regular intervals: say, once a day on busy projects. When users log-in, most systems create a project summary or home page (also variously described as 'headlines', 'dashboard' or 'inbox' – see Figures 5.6 and 5.7, for example) unique to that user which highlights any new drawings or documents sent to them, plus any where comments have been made that they should review. As such, it can be considered as a kind of 'in-tray'. Through this interface (usually designed to be as intuitive to use as possible), they can then click on these items to 'pull' them into their browser window. As mentioned, if a user has not logged-in for a while, most systems have a 'push' option whereby they are sent an email notifying them that something has been published which needs their attention; such emails typically include a link to the user's home page, where he/she can then 'pull' the item into their browser as usual. In some systems, the home page may be configurable, with the user able to specify additional modules of information to appear below or around the project summary. Options might include a project diary, weather forecasts for the site location, or the latest site progress photographs.
- *File viewing* By clicking on an issue notification, recipients will then be able to see who has issued something to them and what, if any, actions (approval, comment, authorisation, etc.) are required of them. They will then be able to view the file on-screen. Assuming that a team has agreed to use standard Microsoft Office tools for word-processing, spreadsheets, presentations and the like, they will be able to open such files (usually on a read-only basis) using the native applications on their own local machines, while file viewing utilities may be needed for other formats (e.g. see Section 5.4 for more on sharing CAD files).

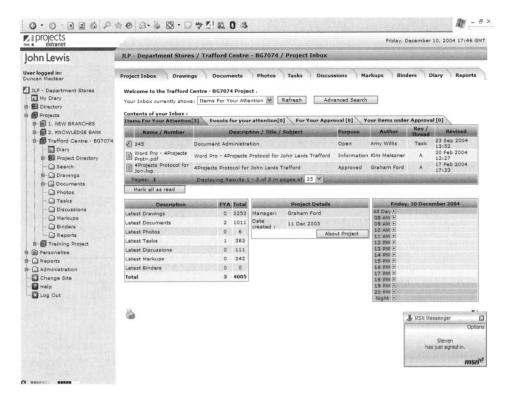

Figure 5.6 Example of project inbox (1) – from 4Projects extranet (by kind permission of 4Projects Ltd and John Lewis Partnership; image copyright to 4Projects Ltd).

- *File history* The construction collaboration technology should detail the history of a particular file: particularly relevant where, say, a drawing has undergone a series of revisions. While the system's default setting may be to show viewers only the latest version, users may occasionally want to review previous issues. Project managers may want to see who has accessed and viewed a particular file, and most systems provide reporting tools that detail who accessed a file, when and what, if any, action (e.g. view only, comment or mark-up, etc.) was taken.
- *File download* Where a recipient wants to edit a document, it will normally first be necessary to download the file onto his or her local machine. Instigating a file download can be as simple as clicking on an icon or filename on the desktop, or, if several files are needed, most systems provide batch download facilities.
- *File edit* It is important to remember that most collaboration platforms will *not* allow end-users to edit a file published to the system and then save the changes as though they were in the original document under the same filename and version. Any changes made will automatically be saved in a new version of the file along with the editor's details and the time and date, etc., helping to safeguard both the integrity of the original files and the system's audit trail.
- *File deletion* Perhaps a misnomer on this list as, like file editing, end-users should *not* be able to delete files from the system, either permanently or by

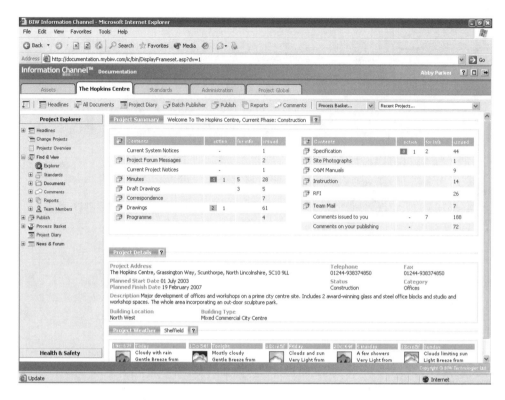

Figure 5.7 Example of project inbox (2) – from BIW Technologies (copyright to and by kind permission of BIW Technologies Ltd).

overwriting them with new revisions under the old filename and version number, etc. (any file deletion, of course, immediately undermines the effectiveness of an audit trail). On some collaboration systems, authorised users may 'disable' files so that they disappear from the end-users' screens, but they remain in the system and, if necessary, could be 'enabled'.

• *File printing* Viewing a drawing or other large document on screen can be difficult. The display may show the full extent of a large A1 drawing, for example, but the drawing's details will often difficult to discern unless the end-user has a very large monitor screen or is perhaps able to project the image onto a wall. Equally, showing the same drawing at a comfortable scale will mean zooming in to a small portion, and losing the 'big picture'. Accordingly, some recipients may wish to print out the drawing (see also 'Commenting and mark-up', and Section 5.4).[8] While this is a core feature of all the leading construction collaboration systems, it is worth checking that the application can print accurately and to scale from the drawing formats supported and can faithfully deliver appropriate line thicknesses, fonts, etc. Most of the systems also support batch printing facilities, allowing end-users to run off prints of multiple files at selected sizes on specified devices, etc.[9]

5.2.3 Feedback

- *Commenting and mark-up* All the leading construction collaboration systems provide tools that allow end-users to comment upon and mark-up or 'red-line' a drawing or other document file. Where the file is a drawing, such comments and mark-ups are usually enabled by tools within a viewer application (see Section 5.4 for a more detailed discussion of viewer technologies) which records the end-user's comments, etc. as metadata in an overlay associated with the drawing file.[10] These comments, etc. are then automatically relayed back to the person who issued the file, along with anyone else included on the issue list.
- *Measuring tools* Some viewer applications allow the recipient to measure distances and areas represented within the drawing for cost estimation purposes.
- *Commenting review* As comments and mark-ups on an initial drawing will normally lead to the issue of a new version, most of the collaboration platforms allow the end-user to review past comments on earlier revisions and relate them to the current revision, allowing them to track the discussion that has taken place about a particular issue or area of the design.
- *Status change* Drawings (and other documents) are often issued for approval or authorisation purposes, and, as well as details about the drawing itself, the collaboration systems may show its status. In some applications, authorised recipients have a 'status stamp' tool that allows them to stamp the drawing as, say, 'approved for construction'. Such status changes will then be communicated back to others on the issue list, and will be recorded in the drawing's audit trail.
- *Discussion forums* Outside of commenting capabilities related to specific drawings or other documents, most collaboration systems offer some kind of threaded discussion forum or bulletin board feature, allowing project team members to, for example, ask for comments on a particular topic, and for other team members to monitor and respond to any of the points made.

5.3 Management features

Many of the features already described are used to support key project processes, helping individual professionals to manage both their own responsibilities and the deliverables required of them. Again, the following subsections outline some of the typical features available:

5.3.1 Workflow management

In an IT context, many of these activities can be described as 'workflows': the automation of a business process, in whole or part, during which documents, information or tasks are passed from one participant to another for action, according to a set of procedural rules.

Across many industry sectors, teams and enterprises often seek to automate standard business processes and traditional 'workflow' approaches are fine so long as everyone adopts the same, often sequential, approach. But in the construction sector, common processes (sometimes best understood as paper-based forms) such as RFIs, technical queries (TQs), instructions and transmittals often vary between different companies and projects, they may vary in their complexity (one simple RFI may be quickly resolved whereas another may take months and involve numerous stages to complete), and inputs will usually be required from across an ever-changing network of individuals and organisations. End-users may prefer systems that do not force a team to use a non-standard 'one-size-fits-all' approach, but allow such forms to be re-configured and even branded to suit their firm's, their profession's or their project's precise requirements.

The range of processes that might be covered include

- *Query management* For example, client queries, TQs, RFIs (including pre-contract, construction, and design RFI variants). Workflows are often configured to suit particular job roles and responsibilities. For example, normally only authorised users may raise an RFI, and there will be rules governing resolution, review, information and forwarding. For instance, someone may initiate an RFI by issuing a request to an authorising controller, asking for a reply within seven days. The controller then passes this request to one or more recipients, specifying the time limit and requesting a reasoned reply. Responses are sent back to the controller who can decide which replies can be forwarded to the originator. In this example, both the originator and the controller may then close the request, and the audit trail of the whole RFI process can be reviewed. Some of the more sophisticated collaboration systems allow these processes to be 'tweaked' to match different companies' existing paper-based processes.
- *Document management* For example, transmittals, submittals, instructions (including variants such as those emanating from a construction manager or an architect). Many project team members use submittals and transmittals to record the transmission of information to a third party. Such forms might, for example, accompany a parcel of drawings, and will record details of those drawings along with the date, sender, recipient, etc. Construction collaboration systems adopt a similar process, and allow such packages to be viewed, reassigned, updated, resubmitted, etc.
- *Change management* For example, early warning notices, change order requests, change orders, change notices and change reports.
- *Approval management* For example, document or drawing approval processes, invoices, etc.
- *Discussions/messaging* Beyond a standard discussion forum or bulletin board system, some teams may want email-like functionality within their collaboration system to record project-related discussions or to pass messages between team members.
- *Print management* For example, print request, print order.
- *Quality management* For example, 'snagging' lists (detailing faults – 'snags' – that need to be rectified).
- *Meeting minutes* On most collaboration systems, meeting minutes and notes of telephone conversations are stored as Word files. Some allow actions and discussion items to be tracked and monitored.

5.3.2   Team management

- *Programme management* Some collaboration systems allow close integration with scheduling or programming applications, and allow project managers, for example, to create, manage and communicate key milestone dates, key deliverables and key actions to specified groups of users. These may also be capable of being added to user's online project diaries.
- *Project diary or calendar* Most collaboration systems offer a project-specific diary or calendar feature, some of which can be synchronised with the user's own diary if it is held in an application such as Microsoft Outlook. Users will be able to view other users' calendars and create, accept or decline invitations to meetings (e.g. see Figure 5.8). For example, the meeting instigator can select the date, time, meeting type and venue (from a list of pre-entered venues, or by adding a new location), and then select attendees. The instigator may also have the option to send an email to all attendees, though the proposed meeting will automatically appear in an attendee's 'headlines'. Links may also be possible to workflows (e.g. RFIs) and to project programme milestones.
- *Reporting tools* These allows users to generate various reports regarding access to the system, document publication schedules, workflows, etc. For example,

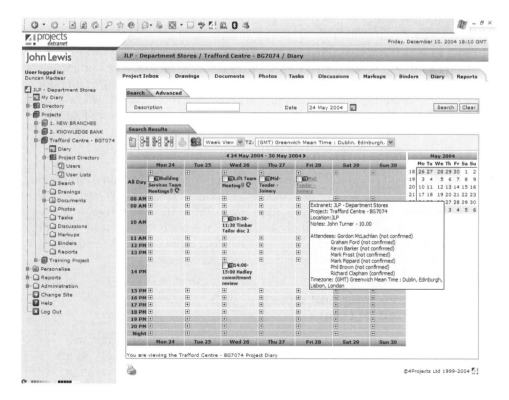

Figure 5.8 Example of project diary – from 4Projects extranet (by kind permission of 4Projects Ltd and John Lewis Partnership; image copyright to 4Projects Ltd).

a user may want to discover who has yet to access a recently distributed new document. A search can be instigated, generating a list of individuals still to open that particular document; some systems will then offer the facility to send a reminder email to each individual listed. Many of the systems allow such reports to be customised and to be exported to other corporate applications.

5.3.3 Work package management

* Most of the collaboration systems support the construction industry concept of work packages: sets of data associated, for example, a particular trade contractor's inputs to a project. This may be achieved by creating different folders for each work package, but this route can become difficult to administer, particularly where files may be shared between multiple packages. Instead, a more sophisticated solution is to publish the information once, but enable that information can be associated with or shared between different packages. The trade contractor is thereafter able to view all the documents, drawings, images, queries, actions, etc. relevant to that specific work package.

5.3.4 Multiple project management

* Where a user is involved in several concurrent projects all of which are supported using a particular collaboration platform, the system will usually provide an easy way to switch between different projects without having to log-out of one and then log-in to another. A single log-in gives access to all that user's projects. This is particularly useful for clients or programme managers overseeing or monitoring a series of projects; some systems provide reporting tools that will allow them to assess the progress of whole programmes of work against various milestones.

5.3.5 Standards management

* Many organisations need to share corporate standards (e.g. branding information, standard contracts, specifications, etc.) across more than one project. Several of the collaboration technology providers support this. In some instances, a central repository of such data is simply created in a project-type environment; other providers manage standards more dynamically, relating centrally held standards with each project-based implementation, and vice versa.

5.4 Sharing, viewing and working with CAD-based drawings

One of the major differences between the construction industry and many other sectors that have begun to embrace collaboration is its extensive use of graphical information. Where other enterprises may exchange numerous revisions of written documents and spreadsheets, for many members of a construction project team – from the client and conceptual designers down to foremen and operatives on site – key information will tend to be relayed in the form of drawings, supported by written specifications. Of course, this is true for many other users of design information (e.g. in manufacturing industries, aerospace, automotive, etc.), but in the construction industry, these drawings usually need to be shared across a widely dispersed group of individuals, many of whom will only have a temporary or short-term involvement with the project in question. While it is relatively simple to relay written information via the web, the core functionality of a construction collaboration system, therefore, must be its ability to manage drawings and all drawing-related processes effectively regardless of the end-users' locations and/or their IT infrastructures.

5.4.1 The CAD file challenge

For hundreds of years, architects and other designers have used drawings and written specification documents to convey information about their ideas so that builders can transform those ideas into reality. Over the past couple of decades, however, there has been a design revolution. Instead of drawing boards, designers now sit in front of computer monitors, editing and manipulating (e.g. erasing, moving, copying, offsetting, rotating, etc.) design information electronically using CAD software packages. Drawings no longer stand alone, but can incorporate associated non-graphical information such as specifications or bills of materials, can be related to other drawings via external references (i.e. X-refs in AutoCAD, reference files in WorkStation), and can incorporate multiple layers of different types of information (see Duyshart 1997; Sun and Howard 2004, pp. 51–61).

CAD outputs were initially still produced in a two-dimensional (2D) paper-based form, which could then be copied and distributed to other designers or forwarded to those responsible for construction. But then designers began to collaborate electronically, sharing CAD file information on disks or CDs, and later as email attachments. Instead of 'reinventing the wheel', recipients could use the information received to synthesise ideas more quickly and efficiently. This, of course, assumed that the sender and recipient used the same kind of CAD software, that the communication channel was capable of conveying information quickly and reliably, and that the CAD files were in a format that allowed data to be viewed and re-used. In many instances, though, different users would be using different CAD packages, trying to send prohibitively large volumes of information, and/or saving data in different formats. Moreover, attempts to impose standards by dictating which software applications and file formats are to be used are almost always doomed to failure; when many project team relationships are temporary and short-term, and software packages may have been acquired at considerable expense, there is little incentive to standardise by switching to another (usually expensive) alternative.

The format of data is critical. Essentially, computer graphics can be defined in two very different ways:

- *Bit map or raster graphics* In a bit map, the colour and display space for each dot or pixel on a screen is defined using a fixed or 'raster graphics' method (a raster is a grid of *x*- and *y*-coordinates). Common bit map formats include TIFF and BMP and, especially on the web, GIF for drawings and JPG for pictures. If such files include a lot of detail or – like most construction-related drawings – are bigger than a computer screen, then the files can be very large (and will not compress very much), leading to large storage requirements and slow network delivery. Such raster files cannot usually be rescaled or modified without losing definition.

- *Vector graphics* Using vector graphics, each line is described mathematically as having a starting point, a direction, and a length. CAD programs generate this type of file, which, while large, are also efficient at holding large amounts of detail and compress well, and so are smaller to store and faster to deliver over a network. Drawing elements can also easily be manipulated on-screen. Most CAD programs generate vector graphics files in proprietary formats (e.g. AutoDesk's DWG, Bentley's DGN). This used to mean that both sender and recipient had to use the same (usually expensive) software, preventing widespread dissemination of the drawing, although increasingly the major packages allow users to open, edit and save files from other widely used applications.[11]

5.4.2   Sharing design information

In the AEC sector, most designers need to share information with, and get feedback from, their clients and other members of project teams. Originally, of course, this involved handing over or sending paper-based drawings and other documents (perhaps by post or courier). In recent years, this has increasingly been accomplished electronically, via floppy disks, CDs, email attachments, FTP sites, LAN or WAN-based EDMSs, and, more recently as this book has described, via internet-based construction collaboration systems. Such methods remove some of the difficulties (and reduce or transfer some of the costs) associated with printing, copying, distribution, storage and retrieval of physical documents. But it not simply a matter of sending the original CAD file. It may not always be possible, necessary or desirable for all recipients to have access to drawings in their original file format. For example:

- not everyone who needs to refer to drawings will have a copy of a particular CAD authoring program (or the appropriate version) – and for every individual on a construction project who requires access to the CAD data in its original, editable form there will be a further four members who require read-only access, while ten won't need access to the drawings at all (Cyon Research 2004, p. 8);
- internet bandwidth constraints may make it difficult to send and/or receive large CAD files;
- designs and design changes will need to be properly controlled and managed (designers may be concerned that layers might be switched off, rendering a design open to misinterpretation, for example);
- there will be issues about protection of intellectual property (designers might distribute documents that could be printed, but not easily copied, perhaps by using 'flat' files or 'static' raster formats).

Disregarding some of the cultural issues and approaches based on forcing every team member to use a standard proprietary software and file formats, some of the technological issues relating to sharing and viewing CAD drawings began to be addressed during the 1990s. Three broad approaches were applied:

- *Page description languages* Beyond the CAD community, software vendors promoted the use of vector graphics-based page description languages (PDLs) to create universal standards for sharing and viewing documents electronically. The two commonest PDLs are Adobe's PostScript language, which later formed the basis for its widely used portable document format (PDF), and Hewlett-Packard's printer control language (PCL).

- *'Universal' CAD file formats* Within the CAD community, some software vendors devised file formats that allowed CAD files to be shared across different applications and operating systems, and to be disseminated efficiently via the web.
- *CAD viewer applications* Also within the CAD community, various CAD drawing viewer applications were developed.

We will now look briefly at each of these.

5.4.3   Page description languages

Since the early 1980s, software developers have examined the potential of vector graphics-based PDLs[12] as a basis for sharing and viewing documents electronically. Among the earliest was Postscript, developed by Adobe in 1985, now an industry standard for printing and imaging (widely used in desktop publishing), and a precursor to Adobe's PDF. The other major PDL is PCL, released by Hewlett-Packard as a faster, simpler and less expensive alternative to Postscript.

- *The PDF approach* Based on Postscript, PDFs are, like CAD files, vector rather than bitmap-based; documents saved as PDF files tend to be smaller than their originals. Widespread availability of the free Adobe Reader has helped the format become the most widely used on the web after HTML, and it provides a common, secure and accessible format for many documents. However, its critics suggest there are some drawbacks to this proprietary standard; for example, they argue (not always convincingly – particularly as software advances often make the claims redundant):

  i   It is not AEC-specific – only once Adobe Acrobat Professional v6 was released in April 2003 did the format supported common AEC drawing conventions such as layering – and then only for AutoCAD and Microsoft Visio.
  ii   Producing PDFs requires most drawing publishers to own the Adobe Acrobat author application or a CAD-to-PDF conversion program.[13]
  iii   Creating a PDF from, say, AutoCAD adds a stage to the publishing process, and can be slow (Autodesk, for example, contrasted the creation of a PDF and a DWF from a DWG, and the latter was created 20 times faster); creation is also a file conversion process and so requires some quality assurance process to ensure integrity; if necessary, re-conversion from PDF back to the original native format is also difficult (although there are applications that allow this).[14]
  iv   Without Acrobat (or other Adobe applications such as PhotoShop or PageMaker), a PDF is something of a 'closed book': it cannot, for example, be easily edited, and – unlike HTML – it cannot easily be linked to other documents (especially if they are in different formats) or information.

    v From a software development point of view, Adobe does not support certain activities (e.g. using an Acrobat product in a multi-threaded way[15] or as a server process accessed by multiple clients); a fragmented, multi-disciplinary project team would therefore need multiple copies of PDF-creation programs to function efficiently – costs which might well be reflected in higher client fees.

    Adobe has also begun to articulate a more advanced vision of the future of the construction industry in which, not surprisingly, its Acrobat product and PDF format feature heavily, with all drawings and other documents being published as intelligent documents (see Cyon Research 2004).

- *The PCL approach* PCL, on the other hand, is not proprietary; it is an open, industry standard, supported by many third-party software and hardware vendors. Designed as a simple interface language, PCL drives printers to produce a true representation of a document or drawing. HPGL – the language used to drive most plotters – is a subset of PCL:

    i third-party PCL viewers tend to be smaller and faster than Adobe Reader, though the latter is free;

    ii PCL files tend to be smaller than PDFs;

    iii while the creation of PCL files also adds an additional stage to the publishing process, the output format is a familiar plot file, not a conversion, and can be accomplished by the CAD package itself (PCL files can also be created by 'print to file' commands from any Windows-based program).

In short, creation of drawing PDFs usually requires an additional application and so there is an additional cost, but they are inexpensive to view via the familiar and free Adobe Reader. PCL files cost little or nothing to produce, may be smaller (saving on network traffic and storage costs), but a recipient will normally need a CAD package, CAD viewer or PCL viewer to review them.

5.4.4   'Universal' CAD file formats

In 1990, Autodesk, developer of one of the leading Windows-based CAD applications, AutoCAD, developed the drawing exchange format (DXF); in 1994, the first vector format specifically designed for the internet – simple vector format (SVF) – was published, and then, in 1998, AutoDesk announced its own new format – DWF – a lighter form of the vector graphics format especially for the web. For organisations sending and receiving lots of AutoCAD design information via the internet (e.g. via email, FTP or a construction collaboration system), the DWF format meant CAD information could be sent in much smaller files (typically less than 10 per cent the size of the original DWG), with a corresponding increase in the speed of transmission. This format was initially of little value to users of other CAD packages (e.g. Bentley's MicroStation, Nemetschek's VectorWorks or ArchiCAD from Graphisoft), but in early 2004 Autodesk – responding to the threat posed by increasingly widespread use of Adobe's PDF format – released its DWF Writer, a free, downloadable application that enabled designers to create DWF files from any CAD or Microsoft Windows application.

One of the major advantages of DWF is its low-bandwidth requirement (a DWF file can be 10 per cent of the size of a PDF file of the same drawing), and – as a result – it has become a popular format for use by the providers of internet-based construction collaboration technologies, usually in conjunction with a CAD file viewer. Being quite compact, it can be also be viewed more quickly than PDF, with panning and zooming much faster processes. (In 2004, Autodesk, as champion of the DWF format, and advocates of the PDF format were engaged in a vigorous debate about the relative merits of their different approaches.)

5.4.5 CAD file viewers

As designers often need to share drawing information with non-CAD users, Autodesk and others launched a range of viewer applications to allow this.

For example, Autodesk's launch of the DWF was accompanied by the release of a free plug-in viewer, Whip! This provided a CAD-like interface within a standard internet browser, allowing users to turn layers on and off, pan around, zoom into areas of interest, and print the drawing off, but it purposely did not provide editing tools, meaning that files could not easily be altered and so safeguarding the integrity of the original drawing. Whip! was superceded in 2003 by Autodesk Express Viewer (renamed Autodesk DWF Viewer in 2004), which offered viewing and printing capabilities, and the more sophisticated – but not free – Volo View, which offers high-fidelity 'read-only' view, plot and mark-up (or 'red-lining') of design data, allowing non-AutoCAD users to track, manage, authorise and/or review DWG, DXF and DWF drawings, plus several raster formats.[16] And the rapid release of new Autodesk products continued in 2004, with the release of DWF Composer in February, which allowed review, mark-up and printing of drawings, and even, within certain constraints, revision of original AutoCAD files.[17]

Fellow CAD vendors have also produced free viewers: Nemetschek launched its VectorWorks Viewer in August 2001, and Bentley launched Bentley View in 2002 allowing users to view MicroStation DGNs as well as AutoCAD DWGs and raster formats. Third party viewers of varying degrees of sophistication (ranging from simple viewing and navigation through to advanced mark-up and measuring tools[18]) and including both free-ware and purchasable applications, have also been developed by, among others, Cimmetry Systems, Informative Graphics, NetGuru, Rasterex and Softsource.

In recent years, such CAD viewer technologies have become a vital tool for the many developers and users of construction collaboration systems. Most UK vendors rely on powerful third-party viewer and mark-up applications. For example, BuildOnline, Causeway and Cadweb use Cimmetry's AutoVue (see Figure 5.9); Asite, Business Collaborator and 4Projects use Brava! from Informative Graphics (see Figure 5.10). BIW developed its own viewer application (see Figure 5.11) offering essentially the same range of viewing, navigation, mark-up, printing to scale and measuring tools, including support for layers and X-refs, but working on a much smaller number of file formats.[19] While using a third-party application may free the provider from a demanding development task, it does mean that the vendor will have to pass on the costs of licensing such applications to its customers and/or end-users; alternatively, end-users may need to license the viewer separately as an additional cost

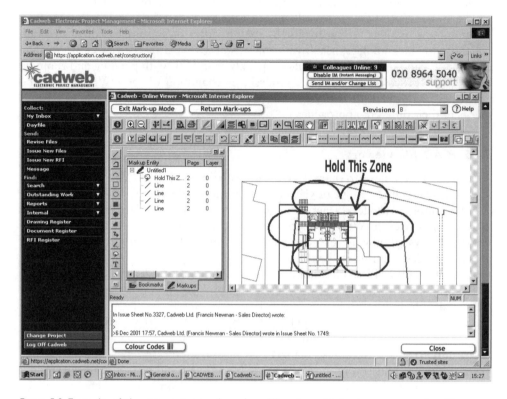

Figure 5.9 Example of drawing review and markup (1) – from Cadweb extranet, using Cimmetry AutoVue viewer (image copyright to and by kind permission of Cadweb Ltd).

of their use of the system. Where vendors are using the same viewer, there will little differentiation between them in terms of viewer capabilities. Vendors will also need to invest in integrating the third-party viewer with their core collaboration system, both of which may be developing at different speeds. In-house development of a viewer application, on the other hand, may mean some compromises on depth and breadth of functionality – albeit at a level unlikely to be discernible to an average construction project user[20] – but gives the vendor greater control over its pricing and integration, and offers additional scope for product differentiation.

5.4.6 *CAD-based drawings in construction collaboration systems*

Given the central importance of graphical information to many construction project team interactions, this section has deliberately been devoted to a detailed consideration of the different methods used to share such information. To sum up: a large volume of project-related documents in the AEC sector are CAD files, but there is, as yet, no ideal method for sharing CAD data. CAD files are produced by a variety of

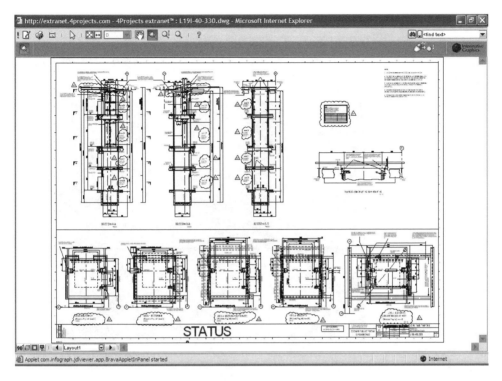

Figure 5.10 Example of drawing review and markup (2) – from 4Projects extranet, using Brava! viewer (image by kind permission of 4Projects Ltd and John Lewis Partnership; image copyright to 4Project Ltd).

different, often incompatible applications; they can be saved in various different formats, some of which can be optimised to reduce file sizes and speed up transmission, if necessary, via the internet. Recipients without the appropriate CAD software either need the right kind of file viewer, or they need the originator to save the drawings in a generic format (i.e. PDF) that can be viewed by a widely available viewer.

So far as the project team's use of construction collaboration technology is concerned, many of the headaches can be avoided by careful consideration of the issues at an early stage, ideally when the project protocol document is being compiled. Most of the technology providers adopt a similar approach to the Project Information Exchange (PIX) Protocol (Building Centre Trust 2004; see also Chapters 7 and 8), encouraging team members to explicitly agree how, in what formats, and with what legal effect, they will exchange documents. Such protocols help to avoid situations where, say, an architect only shares information in 'static' formats in case they may be compromising their intellectual property, etc. Once such decisions have been made, project participants can begin to interact online with the information. So far as CAD files are concerned, with the exception of BIW, this will tend to mean they will be using a combination of the vendor's technology and a third-party file viewing and mark-up application.

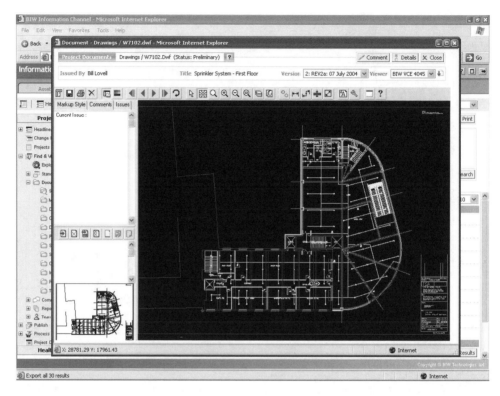

Figure 5.11   Example of drawing review and markup (3) – viewing DWF drawing using BIW Information Channel's integrated viewer (image copyright to and by kind permission of BIW Technologies).

5.5   Construction collaboration technology options

The features and functionality described so far tend to be replicated to a greater or lesser extent in most of the applications available in the UK market today. However, several of the vendors also offer complementary products that either provide additional functionality to the core system, or extend the functionality into new areas.

5.5.1   Tendering tools

Traditionally, tendering tends to be a long, complex, laborious, paper-driven and – above all – expensive process. Even with new technology, many tenders still require compilation and distribution of large volumes of paper (sometimes out-of-date almost as soon as they are sent out), perhaps replaced or augmented by CDs containing a 'snapshot' of all the relevant drawings and documents.

The tender manager will usually need to:

• manage shortlists of potential bidders;
• identify which bidders have accepted the invitation to tender;

- produce and distribute all relevant documentation;
- liaise with, and amend or issue further information to bidders;
- coordinate communications between the project team and bidders;
- ensure confidentiality of each bidder's tender response;
- help relevant project team members evaluate the tender responses;
- notify bidders of the tender process outcome.

Responding to a tender has often traditionally also been a complicated and expensive logistical nightmare. Once potential bidders have received an invitation to tender, they will ideally need to:

- decide whether they wish to receive the tender or not;
- have fast, easy access to all the relevant documentation;
- communicate quickly with the tender manager on queries, etc.;
- ensure the confidentiality of the tender response;
- submit the final tender response(s) (usually with multiple additional copies of the documentation) to the tender manager before the tender deadline;
- find out if the bid has been successful or not.

The tendering stage normally follows a lengthy progression from initial outline proposals to the development of detailed design and contractual documentation. Many organisations use construction collaboration systems to manage the processes *leading up to* the appointment of contractors, subcontractors and suppliers to deliver construction services and materials; they also then use the system to manage the ongoing exchange of information *during and after* the build phase. It makes sense, therefore, to use the same system as a basis for the stage in between: tendering. Several of the leading construction collaboration providers, therefore, offer a tendering option (including 4Projects, BIW, BuildOnline and Sarcophagus). In some instances, the option may be used as a stand-alone application; in others, it may be integrated with the core application so that design information and other data in the collaboration environment is seamlessly re-used for tendering purposes.

The tendering tools all limit bidders' access only to information relevant to the particular tender or work package. Bidders can view, download and print out drawings or specifications as they wish, and can use the system to communicate with the tender manager about queries or changes (with 'workflow' tools used to manage such tender queries, etc.). And, at the end of the tender period, they can submit their responses and all associated documentation direct to the system, secure in the knowledge that only the tender manager can see their submissions.

5.5.2 Webcam facilities

In addition to conventional still images of projects on site, clients and team members may wish to view real-time webcam images of their projects. Some vendors offer this as an option, with further options available. For example, more than one webcam might be set up, or instead of a simple 'static' webcam, the team might be allowed to operate the webcam remotely (e.g. to pan, tilt or zoom-in or zoom-out). The images can be time- and date-stamped, and archived alongside other project data.

5.5.3   Health and Safety File

In the United Kingdom, under the 1994 Construction (Design and Management) Regulations (CDM), the client and project team must produce various health and safety documents during and after construction. One requirement, the Health and Safety File, has traditionally been a substantial, comprehensive and expensive-to-produce library of documents. Collation requires extensive inputs from across the team; often the File can fill over 20 thick ring-binders, and several copies of the whole File may be required. Typically, the File is not completed until weeks after project hand-over. Once compiled, the File is passed to the client, who must then store it, maintain it and make it available to anyone needing information about the facility – whether for routine operation and maintenance or for long-term use, for example: major alteration works. If the client sells all or part of the facility, the File, updated to reflect any further works, must be passed to the new owner. It is, therefore, a key part of a built asset's 'whole-life' documentation.

Building on information already routinely exchanged using its collaboration platform, BIW and Sainsbury's developed a system capable of producing an electronic Health and Safety File which is CDM-compliant, faster, easier and cheaper to compile, maintain and update, and is more accessible to facilities management (FM) staff who need to manage post-construction operation and maintenance processes throughout each built asset's 'whole life' (see Figure 5.12).

Figure 5.12 Example of extranet-based Health and Safety File – from BIW Information Channel: H&S (image copyright to and by kind permission of BIW Technologies).

5.5.4 Project archives

At the end of a project or programme of works (but also, sometimes, at intervals during a project), decisions have to be made about what to do with the body of information created by all the team members. In the past, businesses would maintain paper-based archives of information relating to their own inputs to a project for maybe 15 or 20 years (perhaps to protect themselves against claims and fulfil their professional indemnity insurance (PII) or other insurance obligations after project completion), but much of the unwritten knowledge accumulated during collaborative processes would be lost almost as soon as the team dispersed and its members began working on other projects. In most instances, the client would receive a substantial volume of as-built drawings and other information (e.g. the Health and Safety File), but it would usually be in either a paper-based form (often in numerous ring-binders), or stored on a number of CDs or DVDs.

Construction collaboration technologies offer the client and its project team members a range of alternatives that broadly offer three different levels of sophistication:

- Offline electronic archive as a series of unstructured files (usually on disk) – this would amount to a 'dumb' archive of all the project data, not dissimilar to a simple back-up of all the files by an IT department, but would not include all the metadata (e.g. comments, issue lists, audit trails, etc.). The end-user would also need either the software used to originate different types of document or the appropriate viewing applications to access the documents held on the disk(s).
- Offline electronic archive of collaborative project platform (perhaps on a portable hard-disk) – a self-contained copy of the project data, plus associated metadata and the software applications used to search, access and view that data, is placed on a storage device that can either be stored itself or used to transfer the archive to the end-user's own corporate network.

In both instances, the storage media may hold all the project data (e.g. where it is required by the owner), or it may reflect the permission levels agreed for the project and only hold the project data to which a particular team member had access (e.g. where an archive is required by, say, a consultant, sub-contractor, supplier, etc.).

- 'Live' online archive and FM tool – the collaboration technology provider maintains the data online, allowing the owner and any authorised individuals (e.g. operations staff, maintenance contractors, etc.) to search, access, view and, where appropriate, update the stored information during the working life of the asset or facility.

Particular issues here might concern the durability of the storage media (there have been instances of CDs deteriorating over time, for example) and the continued availability of hardware, software and operating systems used to create or read the stored information. However, in addition to its risk management value, such an archive can become 'an entirely reliable rich source for data mining and post-project analysis and review' (CITSEC 2004).

5.6   Ease of use

To some readers, this chapter may have presented a formidable and even confusing description of the range of features offered by the leading construction collaboration systems. While a prospective customer or project team will want to be sure that a system offers all the functionality that they are likely to need to manage their project or programme of works, they should also pay attention to how easy the system is to use. Some of the early US systems were criticised for not doing more to resolve 'people' issues; as O'Brien (2000, p. 34) wrote: 'project Web sites…[have] not been designed with enough attention to the particular tasks of certain individuals, and this has made it difficult for these individuals to incorporate the tool into their everyday work'. Becerik's survey (2004a) of participants on three US projects found almost 42 per cent of respondents wanted their online tools to be re-designed to make the interface more intuitive and less complex (p. 8).

The web browser-based interface employed by most systems will offer navigation features familiar to any web user (e.g. lickable hypertext links and icons, search engines, etc.), and most of the software providers have gone to considerable lengths to create interfaces that can be configured to match the roles and responsibilities of individual users, and that are as intuitive to use as possible, perhaps with context-sensitive help features to assist users if they get stuck (however it will be clear from some of the screenshots that the look and feel can very greatly between different applications). Users may also be able to customise their own project summary or home page so that it carries as much, or as little, information as they want. Such steps help reduce user-resistance to the system, and also reduce the amount of training required for the average user.

When assessing different systems, customers should not rely solely on a check-list approach to their suitability. The fact that a system possesses a particular feature may not always mean that it is easy to use or that it fits with how a client organisation, a project team member or an individual end-user prefers to work. The number of steps or screens that may be involved in a particular process may also vary between the different systems. The providers should be asked to demonstrate all key system features and show how easily they can be used to manage key project processes. End-users might also be encouraged to try out the systems using their own computers – some vendors have demonstration sites or create 'sandpit' projects where potential users can 'have a play'.

It will also be appropriate to examine to what extent (if any) particular features might be configured so that they accurately replicate familiar forms or procedures (e.g. transmittals, RFIs, change orders, etc.). Some systems offer a rigid, one-size-fits-all approach; others allow customers to 'brand' their collaboration platforms; for example, the initial log-in screen may carry the customer's corporate logo and repeat this on the project summary or home page, while on-screen forms may be modelled on existing paper-based items that will already be familiar to end-users within that organisation. Such steps not only promote ease of use, they can also help achieve user buy-in – rather than regarding it as an outside system, users come to regard it more as *their* system.

Clearly, successfully introducing a construction collaboration system into the project team is only part of the process. Organisations will want to be sure that the system is capable of supporting their future information, communication and business process requirements. A simple, easy to use system may be perfectly adequate if it is only to be used to facilitate file sharing during design and construction of a small project, but a more fully featured system may be required for larger projects and teams, or for multi-project programmes, particularly if the information is to be re-used beyond the completion of individual schemes. In the latter instance, customers might be advised to seek systems and providers that offer the best of both worlds, that is, an easy to use system that can be expanded or extended as the users become more experienced and sophisticated in their needs.

5.7   Support services

On its own, IT, however well designed and configured, will not resolve an organisation's or project team's problems; the technology implications clearly have to be considered alongside all the relevant people and process issues. The provider's consultancy, training and support services, therefore, can make a crucial difference to the successful delivery of web-based, mission-critical construction collaboration applications. Nitithamyong and Skibniewski (2004) identified five potential success factors relating to the service providers (contact facilities, promptness of responses, attitude of staff, technical competence of staff and knowledge of construction business and problems) and found all five had a significant impact on the performance of the service.

Relevant experience and good project management skills are vital. Effective support services will often combine expertise in configuring and using the software application with experience of working on UK construction projects. The leading UK vendors tend to offer first-line support from consultants with construction industry backgrounds working alongside colleagues with complementary software implementation skills, supported by second-line help-desk support staff, with the provider's SLA specifying levels of support service availability (e.g. staff will be contactable during normal UK working hours).

The depth of experience will vary from vendor to vendor: clearly, the more implementations that a vendor has undertaken the more experience its consultants and support staff will be able to draw upon. Such support teams will be able to draw upon proven best practice and to share experiences gained from previous deployments, both with new project teams and, in some cases, with teams working for clients engaged on long-term multi-project programmes.

However, one size does not fit all. The base implementation methodology needs to be adapted to suit each client's requirements and each team's different skill levels and working practices. And interpersonal 'chemistry' between the provider's staff and end-users within the project team will also be an important factor in both achieving a successful implementation and maintaining an amicable working relationship throughout the lifetime of the project.

The basic range of support services may include:

- working with the customer and its supply chain to understand their technology skills and resources and their process requirements, to identify risks, key objectives and desired benefits, and to agree key performance indicators;
- compilation of a project protocol document – perhaps using the PIX Protocol (Building Centre Trust 2004; see Chapter 8), a vendor-specific protocol, or one

developed a project team member – detailing the project processes and deliverables and the technologies and information exchange standards the team will use;

- configuring the collaborative environment so that it is specific to the process needs of each programme/project and/or project team and/or end-users;
- helping identify and resolve any hardware, software or telecommunications issues, e.g. advising end-users on effective connectivity (see Chapter 6), liaising with internal IT departments over IT and internet usage policies, infrastructure issues, etc.;
- training team members in the most appropriate use of the system for their specific roles, responsibilities and requirements (see also Chapter 8);
- commissioning the system and handing it over to the project team;
- providing ongoing face-to-face user support, plus technical support via a helpdesk service (normally offering both telephone – perhaps with free or low-cost call rates – and email routes), throughout project delivery and, where appropriate, into the post-delivery phase.[21]

5.8   Chapter summary

As previous chapters have emphasised, selecting an application is not simply a question of picking the solution with the most features. Decision-makers need to weigh up the credentials of the vendor; they need to look at the hosting options and, if opting for an ASP-delivered solution, they must also weigh up the hosting infrastructure and the credentials of any third parties involved. And when looking at the technical capabilities of each system, they need to be sure that it strikes a balance between meeting the immediate and forecast needs of all project team members, and being as intuitive and easy to use as possible.

This chapter has looked at the features of construction collaboration applications typically employed to meet the practical information and communication needs of a UK construction project team. Given the pace of technological development it has not sought to provide a detailed guide to every area of functionality (such a guide would quickly become out-of-date); instead, it has focused on outlining the principle generic features of the leading systems on the market. It has also emphasised the key role to be played by the vendors' implementation, training and support teams in ensuring that end-users make best use of their chosen system. Part of their consultancy role will usually relate to how end-users might connect to the system – which is now covered in more detail in Chapter 6.

Connecting to a construction collaboration service

This chapter:

- describes the changing telecommunications infrastructure, notably the recent growth in broadband connections;
- explains some of the key variables in achieving an effective connection to an internet-based service;
- outlines some considerations relating to operating systems and browsers.

Growing use of IT solutions within the AEC sector means that more attention has to be paid to communication capabilities, at both the corporate and individual levels. The Construct IT guide 'How to manage e-project information' (2003, p. 10) recommended several benchmarks to assess:

- How many key project persons have access to a computer?
- Are computers grouped within the same building or do users access remotely?
- What is the computer literacy of these persons using computers?
- What key software is installed on computers (CAD versions, Adobe Acrobat Reader, Excel, Access, programme software, etc.)?
- What speed of internet connectivity do they have (e.g. asymmetric digital subscriber line (ADSL) connection or telephone dial-up)?
- Do all the computers have access to a local server?
- Are there any capacity, functionality or performance issues which need to be addressed?

At one time, using a software application usually meant that the user (or perhaps a colleague from his or her IT department) had to install the package so that it could be opened and used from the user's computer, perhaps via a corporate LAN or WAN where appropriate. The increased use of internet-based applications such as construction collaboration technologies, however, means – as the final three bullet points above suggest – that more attention has to be paid to the connection between the end-user's computer (or corporate LAN/WAN) and the internet. Simply making a connection may not be enough; the connection will also need to have adequate capacity to carry data to and from the end-user's machine. This chapter, therefore, focuses on the key variables to consider in achieving an effective connection to, and interacting effectively with, an internet-based solution.

Once seen as the preserve of 'geeks' or 'techies' (IT professionals), the internet has now become an indispensable tool for many businesses and other users. By 2004, the UK telecommunications regulator Ofcom was reporting that more than two-thirds of all the UK businesses and half of all the UK households had internet connections, and a growing proportion were using broadband links. The internet has revolutionised communications, providing new ways to market services and products, and enabling new business processes. Email has, in many instances, replaced traditional post and the fax; company websites are often preferred to brochures; many transactions now take place online; and – in the early years of the twenty-first century – the advent of web services, and of internet-based construction collaboration technology in particular, has begun to change the way many people in the AEC sector work with each other.[1]

6.1   The growth of broadband

By 2004, one in two of the UK population were internet users. The explosive growth in access to, and use of, the internet has both stimulated, and been stimulated by, some significant advances with respect to telecommunications technology. During the early/mid 1990s, for example, early web converts were often relying on – by today's standards – slow external modems (9.6 Kbps or 14.4 Kbps); next came 28.8 Kbps external modems; and by the early 2000s, by contrast, most PCs came with an internal 56 Kbps modem as part of the standard configuration. Today, old-style, traditional copper-based telecommunications lines have been augmented by ISDN, ADSL/symmetric digital subscriber line (SDSL), fibre-optic cables, wireless technologies, satellite, etc. (in August 2004 BT forecast ADSL broadband would be available in more than 99.6 per cent of the UK telephone exchanges by summer 2005 – available to as many homes as good quality TV reception; Ofcom (2004) estimated that ADSL/SDSL was available to 84 per cent of the UK homes and businesses, more than 45 per cent of the UK homes could use cable modems for broadband access to the internet, while 13 per cent could potentially use broadband via fixed wireless access).

The telecommunications and ISPs have also extended their ranges of connection options from simple dial-up narrowband services offered on a metered, 'pay-as-you-go' model, right through to un-metered, 'always-on' broadband access. A growing number of the UK businesses and households now have access to broadband (see Figure 6.1). Ofcom estimated there were approximately four million broadband subscribers as at the end of April 2004, including almost 2.45 million ADSL/SDSL subscribers and around 1.54 million end users of cable modems. Growth was accelerating (up from three million just six months earlier), passing the five million subscribers landmark in September 2004. BT customers alone totalled four million in December 2004. And the total number of broadband subscribers was forecast to reach eight million by the end of 2005 (according to Telecom Markets' Broadband Subscribers Database).

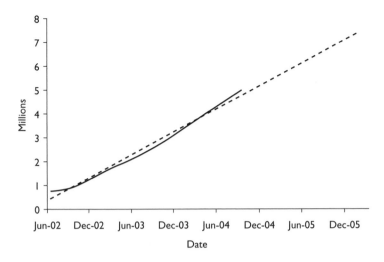

Figure 6.1 Growth in the UK broadband subscribers since June 2002.

6.2   Low bandwidth or not?

Notwithstanding these developments, some of the UK construction collaboration technology providers have adopted a conservative approach to connectivity, trying to make their applications accessible over quite basic telecommunications links. Certainly, around 2000, when many of the UK vendors were first launching their businesses, this approach was born out of two pragmatic realities: first, in a typical construction project, many of the participants might be quite small businesses with limited IT budgets; second, individual project participants often wanted the convenience of being able to access the collaboration system from home, site offices or other locations. In both cases, what works well in an office may not always be available or economically viable in a home or a temporary site cabin, so the lowest common denominator was often a standard telephone dial-up internet connection via an ISP.

Some providers consciously designed their applications to be usable over a standard 56 Kbps modem, particularly where the user's main activities were viewing and commenting upon other team members' drawings and documents. However, where a user's work involved significant amounts of publishing and retrieving of information (i.e. uploading and downloading files to and from the collaboration system), such a connection could prove slow and erratic. In such instances, providers might recommend that the end-user(s) have access to ADSL/SDSL or an ISDN line as minimum.

The key word in the last sentence is 'recommend'. A vendor can design its internet-based construction collaboration application to make the best of the sometimes limited connectivity that exists in European telecommunications infrastructures; if it an ASP, it can ensure that its web services are always available, that they are delivered at high levels of performance from the data centre, and that – up to its connection with the internet – there are no bottlenecks. Obviously, if the application is being hosted by a project team company, then the responsibility of ensuring high levels of performance and availability falls to that business's IT team.

However, in neither instance will either the vendor/ASP or the hosting project team member have any control over how the application will be delivered once it leaves its data centre or IT department (Figure 6.2).

As far as the construction collaboration technology vendor is concerned, the telecommunications infrastructure used by the office-based end-user to connect to the internet is the responsibility of each project team member company (or the individual end-user, if he/she is accessing the application from home or some other location). They should undertake their own sizing exercise to establish what line type and speed would be best suited to their organisation and business needs.

In some instances, the main contractor may also be responsible for providing internet connectivity to the project site for all site-based team members, perhaps asking sub-contractors to include a sum of money within their tender for provision of this service, particularly where it may also be used for their email and other corporate needs. Mindful that the site will usually be visited by non-site-based team members, the contractor may also provide computer terminals and printing facilities that can be used by visitors.

Assuming users also have access to the internet for other purposes (e.g. email) and that they place average demands on their chosen collaboration system (including upload and download of drawings as well as viewing), providers recommend various higher-capacity connections for different levels of multiple usage of their system; for example, Table 6.1.

However, whether users are on large, secure corporate systems or on small home set-ups, access problems can still arise due to problems with:

- the local connection between a user's machine and his/her company or site network;
- the connection between the company or site network (or the end-user's stand-alone machine) and the internet itself.

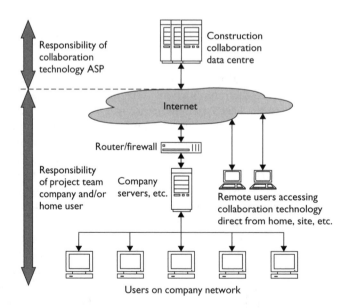

Figure 6.2 ASP and customer connectivity responsibilities (simplified).

Table 6.1 Recommended connections for different numbers of users

Number of users	Type and speed of connection
1–20	128 k ISDN
20–60	256 k leased line
60–150	512 k leased line
150–500	1 Mb leased line
500–3000	2 Mb leased line
3000+	4 Mb leased line

If and when connection problems relating to use of a construction collaboration system do arise, in the vast majority of cases it usually proves to be some problem with the infrastructure used to access the internet, either within the work or home environment, or relating to the telecommunications provider or ISP which supplies the link to the internet. Such problems can usually be avoided or rectified by careful consideration of the relevant infrastructure, either at the company level or at the home or remote user level.[2]

6.3 Internet connection via a company or site network

Research by BT Openworld (2002) reported that the UK's construction firms were lagging behind their counterparts in other industries in upgrading their internet connections to cope with large amounts of information. A survey of 100 small and medium-sized businesses revealed that over 50 per cent still handled business using paper and did not have enough bandwidth to receive electronic files, despite working with CAD drawings, plans and legal contracts.

The BT survey did, however, identify that architects were then the industry's front-runners in investing in broadband, while 'feedback from quantity surveyors, smaller contractors and engineers showed a reluctance to work from soft copies, preferring to rely on post.' A BT Openworld spokesman believed industry businesses needed to evangelise the benefits of broadband to their partners and suppliers: 'After all, there's no point in having high-bandwidth connections if you can only send fax to most recipients', he said.

6.3.1 Selecting an ISP

Above all, companies should make sure that their chosen ISP has a good track record and that it can deliver what it says it can deliver. Buyers should seek evidence of past performance and ask for details of current clients for reference purposes, and the lowest-cost providers may not always be the best.

A guaranteed minimum level of service is essential (some ISPs can choose inconvenient times to undertake network maintenance). It is usual when buying leased lines (also sometimes known as dedicated lines) that a back-up line (perhaps ISDN or a duplicate of the primary line) is also provided; this is essential if connection to the internet is vital to the business.

If current bandwidth requirements are known but the company might need to upgrade in the near future, it should consider installing a large line but have it 'throttled' so that the company only uses and pays for the smaller bandwidth until the upgrade is initiated (for instance, the company could purchase a 2 Mb line and have it throttled to 512 Kb and so only pay for the usage of a 512 Kb line). Once the upgrade is requested, most ISPs can achieve this within 24 hours.

The ISP should be obliged to provide regular statistics on utilisation of the leased line so that the business can plan for upgrade(s) before they become essential. For corporate use, the selected ISP should not share its network with domestic users, as internet performance may be degraded by heavy usage by non-corporate users.

6.3.2 Internet gateway configuration

Figure 6.3 shows a typical secure corporate internet gateway and is based on a user population of around 1500 medium use users (NB: this is only an example and may not be appropriate for all networks and for a small organisation, see Figure 6.4).

This represents a resilient internet gateway that could be deployed on a corporate network. Its internet connections have automated fail-over should the primary line

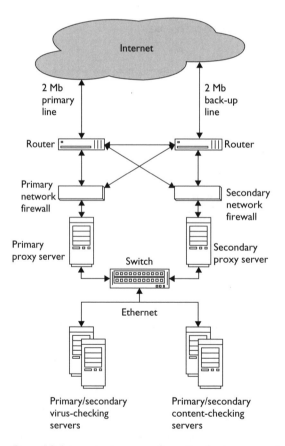

Figure 6.3 Internet gateway configuration for large organisation.

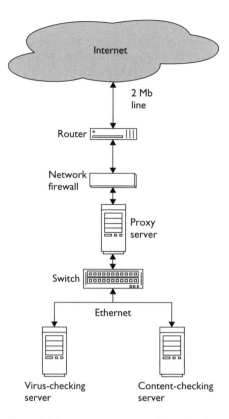

Figure 6.4 Internet gateway configuration for small organisation.

fail, and there are dual firewalls (with a primary on-line and secondary in standby mode) with automated fail-over should the primary fail. Depending on their internal security policies, some companies might also employ two levels of firewall for added security. The diagram also includes resilient content filters to stop users surfing inappropriate material, and resilient virus checkers which could be used to virus check the corporate email system as well as Internet content. The final element, and the one that users would initially connect to, are resilient proxy servers that can be used to authenticate those users who are authorised to browse the internet as well as caching frequently used web sites.

6.3.3 Company network – common internet access problems

Sometimes end-users can believe that their chosen construction collaboration technology website is no longer accessible when, in fact, there can be a more general problem with access to the internet as a whole. For example, there may be instances when a particular website becomes so busy that it either delivers pages very slowly or fails to deliver pages at all.[3] In such cases, the presence of a fast internet connection would have made no difference.

In most normal circumstances, however, there are various possible causes for no internet access or low speed access. Users on a company network, for example, may experience slow access due to poor internet gateway configurations:

- *Inadequate line size for internet use* For example, if a company employed 200 staff, most of whom regularly accessed the internet during their working day, and used a 128 Kbps ISDN line to the internet, they would experience slow access (and occasionally no access), due to the inadequate size of the line. An upgrade would be recommended in such cases. Similarly, some organisations find that more employees access the web at lunchtime (perhaps surfing the web, booking holidays or checking their bank balances while eating their sandwiches at their desks), with a temporary deterioration in internet access until employees return to their normal working patterns. Again, either an upgrade could be implemented, or new constraints imposed on when and how staff can access and use the internet (see discussion of contention ratios in Section 6.4.3).[4]
- *Under-performing ISP* Most businesses access the internet via an ISP. A huge internet 'pipe' may be in place, but access may still be slow. One cause may be the ISP's own network. In such instances, the business should ask its ISP to provide bandwidth and access speed statistics so that it can monitor the ISP's performance.
- *No proxy server* Although not always required, a proxy server can speed up internet access quite dramatically. Regularly accessed web pages are stored locally on a proxy server so that users access them from within their internal network rather than downloading them from the internet each time.
- *Tiered network infrastructure* In many companies, staff may be located in more than one office. The central or head office may host the main internet gateway with access to the internet via a substantial 'pipe'. But rather than having direct internet access, regional or subsidiary offices may rely upon the head office network and 'pipe', to which they have access via slower lines (e.g. via ISDN). Internet traffic from such local offices will often be slower as it has to negotiate the ISDN line before it actually reaches the main internet gateway.
- *Content filter installed on inadequate hardware* Some internet gateways have content filtering software installed to stop users from surfing for inappropriate material from the web. If the software is installed on a low specification machine with low memory and low speed processor, it may become a bottleneck. Content filtering is demanding in terms of memory and processor speed requirements.[5]
- *Virus-checking software on inadequate hardware* Similarly, although not as resource-hungry as content filtering software, virus-checkers still need adequate hardware to perform without degrading service performance.
- *Over-complicated rule set on firewall* Firewalls are an important element of an internet gateway infrastructure, but if their rule sets are over-complicated this can slow down the speed at which traffic flows through them. Though unusual, this possibility may sometimes arise.
- *Proxy, firewall, content filter and virus checker on same server* Sometimes, a network administrator will recognise the need for proxies, firewalls, content filters and virus-checkers but install them all on the same server, causing a huge bottleneck because each function requires considerable system resources.

- *Inadequate internal network* If all the above issues are addressed, the Internet gateway may be perfectly adequate, but if the internal network is poorly set-up users may still experience problems in accessing the gateway. Too many network hops, poorly configured routers or inadequate hardware can all cause problems.

6.4   Internet connection for remote users

6.4.1   Selecting an ISP for remote users

As with selecting an ISP for corporate use, the cheapest is not always the best. Free dial-up accounts, for example, may appear attractive, but are more likely to have slow access times, expensive telephone support services, etc. (if a free account is needed, the best ones are generally those that are less well known). Some ISPs also impose limits on how much data can be downloaded over a period (which may be a problem if the end-user needs to access lots of drawings or other documents). The better ISPs are those that offer business accounts without download limits and guarantee that no free users have access to their network, giving faster access times and a more reliable service.

6.4.2   ADSL/SDSL

Recent years have seen dramatic growth in the spread of broadband communications, notably through ADSL or SDSL. Where the end-users of construction collaboration technology want markedly improved performance but ISDN is not feasible or afford-able, ADSL can help boost productivity and prices charged have dropped considerably in recent years. Based on a 512 Kbps service, for those users who mainly need to download drawings, ADSL is typically up to 11 times faster than a traditional dial-up modem; when uploading drawings, the service is usually up to five times faster than a modem connection.

Users situated close to local telephone exchanges may have access to faster ADSL services: 1 Mbps (normally available within 6 km of an exchange) or 2 Mbps (available if users are within 4 km of an exchange). SDSL has been slower to roll out across the UK telephone network. In August 2004, SDSL at speeds of up to 2 Mbps was only available at 184 exchanges (with a further 120 exchanges to be added), mainly in urban locations, but 'local loop unbundling' – where an ISP can install its own equip-ment in a BT exchange to allow faster services – may expand the network still further in the months and years ahead. In some city locations, even faster services – up to 4 Mbps – began to be marketed during 2004.

6.4.3   Contention ratios

However, particularly where users might be considering a broadband connection via ADSL/SDSL, a key issue to explore is what contention ratio is offered by the ISP. This indicates how many people might be sharing the same internet connection line simul-taneously. In short, the more people there are fighting for the same internet access, then the slower the connection to the internet can be (if there are 20 people to one line, then the contention ratio will be 20:1; if there are 50 people to one line, then the contention ratio will be 50:1 – a contention ratio of 20:1 will give faster access than a 50:1 contention ratio).

In the United Kingdom, BT has set maximum contention ratios for ADSL of 20:1 for business users and 50:1 for home users (it offers a 10:1 maximum contention ratio for its SDSL service). Of course, the actual ratio achieved will vary at different times of the day and on different days of the week,[6] but looking for an ISP offering a business standard service may prove beneficial for many end-users.

6.4.4   Remote user – common internet access problems

Once again, sometimes a perceived problem with access to an individual website is actually a more general problem with a user's access to the internet as a whole. If problems persist, then there can be a variety of reasons for this; for example:

- *Slow modem* For some users, once internet access is achieved, poor performance must be the fault of the website, not their hardware. However, any user relying on, say, a 28 Kbps modem to download a 2 MB file will find it takes a long time to download compared to the performance of a corporate network. While education can help such users to appreciate the technical limits of their hardware, it may be appropriate for them to look at upgrading their modem to a 56 Kbps model or switch to an ADSL/SDSL connection (if locally available).
- *Under-performing ISP* Just as mentioned in respect of corporate network users, a poor ISP can cause degradation in internet access speeds (ISPs that offer free access and free gimmicks are often the slowest). As they attract large numbers of users, these ISPs' networks will become heavily utilised, reducing speed for all customers. Users should select an ISP based on performance rather than price, and – as mentioned – consider those which offer services specifically for business users only.
- *Low specification local telephone exchange* Speed of internet access will also depend on the availability within the local telephone exchange of modern telecommunications equipment (e.g. users with a 56 Kbps modem may achieve a 44 Kbps connection in London but only 29 Kbps in a remote rural location). There is little that can be done about this (apart from move house!), though the expansion of access to 512 Kbps ADSL to 99.6 per cent of the UK population means only a tiny minority of users will be so affected.
- *Firewalls, anti-virus, content-checking, etc.* Particularly with the increasing residential use of 'always on' ADSL/SDSL connections, experts are advising users to install firewall software on their home machines or on any laptop that may also be used at home (Zone Lab's ZoneAlarms is a widely used, free firewall application in the home PC market), along with up-to-date anti-virus software (e.g. Symantec's Norton AntiVirus). These applications (and any content-checking tools used, say, to ensure children's 'safe surfing') may also affect internet access speeds.

6.5   Operating systems

Given the widespread dominance of the Microsoft Windows platform – used on more than 90 per cent of all home and office computer desktops – it is no surprise to find that the main construction collaboration systems are all designed to support users working on Microsoft Windows desktop operating systems. The precise scope of product support will vary from vendor to vendor (some will extend back to Windows 98, others may focus on Windows 2000 and later), and project team users should highlight any potential issues with the vendor at the earliest possible opportunity – perhaps when compiling the initial project protocol documents.

However, we should not forget the substantial minority of computer users – including architects and other creative professionals – who have chosen to use Apple Mac computers for their superior graphic capabilities. While some surveys[7] suggest the Mac OS is employed on less than about five per cent of all computer desktops, use within the architect community is appreciably higher (a survey of 60 UK architectural practices by Fedeski (2002) found Apple Macs were employed in 18 per cent of practices, with a further 12 per cent using Apple Macs and PCs; the Barbour Report 2001 found 11 per cent of architects were Apple Mac only, while a further 9 per cent used Macs and PCs). The basic user interface of most construction collaboration systems can be accessed by an Apple Mac user using a standard web browser, but if that user then wants to open, view and perhaps comment upon drawings, most of the common viewer technologies do not work on the Mac OS. Given that the communication needs of a small but nonetheless influential group of users have to be supported, the usual route has been for architects to use an emulation package such as Microsoft's Virtual PC which allows Mac users to run Windows applications, connect to PC networks and share files with PC users from their Apple Macs. However, the author understands that at least one major provider was planning to release a version of its collaboration technology that worked on PC and Mac platforms without emulation.

While not strictly an operating system problem, some corporate users may still experience difficulties in accessing the viewer plug-in technology if their organisation 'locks down' their computer configuration. Desktop lockdown is a step that may be taken by IT administrators, for example, to prevent unauthorised installation of additional software applications (either by the user or as a result of some malicious virus or hacking attack), usually for security or software licensing reasons. Where organisations apply such a policy, it may be necessary for the construction collaboration technology vendor to negotiate with the IT department and get the viewer added to the organisation's 'white list' of permissible applications – at least, for those staff whose work will require it.

6.6 Browser support

In the mid-1990s, the most widely used web browser was Netscape Navigator, which in 1997 commanded around 72 per cent of the market before the release of Microsoft's Internet Explorer version 4. At the time, most websites (and therefore any web-based collaboration systems) tended to be developed to work best using the Netscape browser, but the advent of IE4 began to change the market. IE4 offered many new features and was marketed very aggressively (perhaps too aggressively – leading to Microsoft's prosecutions for antitrust violations), being bundled and pre-installed with Microsoft's Windows 98 and subsequent operating systems on the majority of PCs. The leading collaboration technology providers responded, first, by optimising their systems to support both browsers. However, Netscape's market share plummeted; by early 2004, IE was being used by around 94 per cent of all web users. Some vendors continued to support users of Netscape Navigator, perhaps stressing version 4.6 or later; others dropped references to Netscape in their support documentation, perhaps recommending use of Microsoft IE6, with service pack 1.

With no new significant Microsoft browser development since 2001, users become critical of some of IEs shortcomings (e.g. slowness, no blocker for pop-up windows, no built-in search, poor security), other developers began to incorporate new functionality into competing browser products (e.g. Opera, Mozilla's open source-based Firefox and Camino, and – expressly for Apple Mac users – Apple's own Safari browser), and during 2004 some of these began to chip away at Microsoft's domination of the browser market.[8] Few collaboration technology vendors recommended these products (although any vendor delivering a Mac-compatible system will presumably recommend the Safari browser), but the situation may well change if and when other browsers do begin to significantly threaten Microsoft's market share.

6.7 Chapter summary

As mentioned briefly at the end of Chapter 5, many of the potential hardware, software and telecommunication obstacles may be addressed when the customer and its supply chain start talking to the collaboration technology vendor's staff about their technology skills and resources. Consultants and help-desk staff in the more experienced vendors have a wealth of detailed knowledge of the performance of their application on a wide range of different computers, networks, operating systems and browsers, and can help individual organisations and users identify performance issues relating to their infrastructure and their telecommunications link to the internet.

Resolving any obstacles or bottlenecks experienced by an individual supply chain member, however, is not the responsibility of the vendor. Neither, in most cases, will the supply chain member be able to charge the full cost of upgrading its infrastructure (moving from dial-up internet access to ADSL, for example) to the project. In some instances, the project may bear part of the cost, recognising that the step is necessary for the company's involvement in the project (and to promote use of the collaboration system across the project team), but it could reasonably be argued that the upgrade will have wider and more long-term benefits to that company beyond the project in question (i.e. the organisation's use of construction collaboration technology may simply have accelerated a system or telecommunications upgrade that would eventually have happened whatever the case).

The UK construction industry is in the midst of a technology revolution. The internet has become an indispensable tool for many AEC businesses and staff, and in the process has helped stimulate massive growth in the demand for reliable, faster and higher capacity internet connections. As this chapter has emphasised, internet-based construction collaboration systems may add to that demand, but the providers can only make recommendations to individual organisations and end-users. It is not enough to evaluate the merits of a particular provider and its application, a customer and its supply chain will need to look closely at its own technical resources so that it can ensure it has adequate internet connectivity. But IT-related expertise is not the only new area of learning required. Successful implementation will require careful attention to the legal issues and to the human dimension of introducing and using new technologies. These issues are covered in Chapters 7 and 8.

Chapter 7

Legal issues relating to construction collaboration technology[1]

This chapter:

- briefly considers the legal status of electronic communications;
- describes how to manage contractual relationships with software providers, including 'worse case scenarios' relating to interruption or termination of services;
- extends discussion to include audit trails, e-discovery, copyright, insurance and archiving issues.

Many thought partnering might reduce the role of lawyers in the construction process, but the AEC sector is notoriously risk-averse, and experience suggests few (if any) UK organisations are yet prepared to procure work on trust alone. This probably applies doubly when clients and their project teams are considering using new IT and need to form partnerships with, as yet, still relatively unknown collaboration technology business partners. Clients and end-user businesses have, accordingly, sought to protect themselves through contracts and other legal documents governing their working relationships with their chosen collaboration vendors.

This is no easy task given that collaboration technology is relatively new to many parts of the AEC industry. Construction and technology lawyers have only slowly begun to tackle some of the issues. Indeed, notwithstanding some of the cultural and practical issues about using the technology, concern about legal issues has been cited (e.g. Goodwin 2001) as a reason why wider introduction of such IT systems has been delayed within the UK industry leaving the vendors and their legal advisors to pick their way carefully through a still-evolving legal landscape. However, legal articles in AEC industry publications on both sides of the Atlantic (e.g. in the United Kingdom: Birkby and Nugent 2002, CITSEC 2004, Hampton 2001; in Ireland: McBride 2003; and in the United States: Berning and Diveley-Coyne 2000, Berning and Flanagan 2003, Berning and Ralls 2002, Joch 2002) have begun to highlight a few areas of particular legal concern, including:

- the legal status of electronic communications;
- legal relationships with the software vendors (particularly if they are functioning as ASPs;

- reliability, security, service interruption (including possible loss of information) or unforeseen termination of the service;
- audit trail issues, particularly those relating to discovery (the process of releasing copies of all relevant documents and correspondence relating to a court case) and – where relevant – open government.

This chapter looks in turn at each of these areas. It also briefly examines issues relating to copyright, and the archiving of information at the end of a project.

7.1   Legal status of electronic communications

Broadly, the same legal principles apply whether parties communicate on paper or electronically. Paper-based records (documents, drawings, etc.) are simply records kept in a particular medium, and electronic media are no less valid. But it is important to ensure that records are authentic, accurate and accessible.

7.1.1   Contract provisions regarding electronic communications

Although the use of electronic media to exchange information within the AEC industry has become progressively more widespread (from email, through floppy disks and CD-ROMs, to FTP sites and now project extranets), concerns about the legal status of electronic communications did cause some AEC professionals to delay decisions about using collaboration systems. This uncertainty was compounded by the absence of appropriate provisions in many standard contract forms. Few made explicit reference to the possible use of ICT; important communications were assumed to be delivered 'in writing' in a tangible, paper-based form, such as by a signed contract, by letter or fax or by the issue of drawings.

For example, many construction contracts require participants to issue formal notices, but are not always clear about how such notices shall be given and whether electronic communications will suffice. In a presentation to industry professionals at the Institution of Civil Engineers in London in December 2001, solicitor Ed White of Masons highlighted the relevant clauses of three standard contracts:

JCT

Where the contract does not specifically state the manner of giving such notices, such notices shall be given or served by any effective means to any agreed address.

Engineering and construction contract

13.1   Each instruction, certificate, submission, proposal, record, acceptance, notification and reply which this contract requires is communicated in a form which can be read, copied and recorded.

13.2   A communication has effect when it is received at the last address notified by the recipient for receiving communications or, if none is notified, at the address of the recipient stated in the Contract Data.

FIDIC

1.3 in writing and delivered by hand (against receipt), sent by mail or courier, or transmitted using any of the agreed systems of electronic transmission as stated in the Appendix to Tender...

In White's view, the first two contracts needed amendment to be appropriate for projects where an online collaboration system is employed. He cited with approval some sample clauses from a retailer's contracts:

1.1 'Writing' includes e-mail, facsimile transmission and/or communication in another durable medium that is available and accessible;

2.1 Any communication sent electronically by e-mail or otherwise:

2.1.1 Will be deemed to have been sent once it enters an information system outside the control of the originator of the message.

2.1.2 Will be deemed to have been received by the intended recipient at the time that in a readable form it enters an information system that is capable of access by the intended recipient.

A paper 'The role of electronic information in construction contracts' written by solicitors Hammond Suddards Edge (2001) outlined standards and rules that can be applied to facilitate the use of electronic communications, including:

UK Standard Interchange Agreement – provides a standard protocol for use by parties who wish to communicate electronically. It mainly concerns security, such as authenticity and integrity of data, when it was received and how data logs should be stored and maintained. The Joint Contracts Tribunal provides, as an option, for the provisions of the Standard Interchange Agreement[2] to be incorporated into construction contracts based upon the JCT's standard form. Where the option is chosen, the contract provides that any communication that must be in writing will be validly exchanged when sent electronically.[3]

7.1.2 Legal admissibility of electronic communications

White addressed the issue of legal admissibility by reference to the Civil Evidence Act 1995:

Where a statement contained in a document is admissible as evidence in civil proceedings, it may be proved:

(a) by the production of that document, or

(b) whether or not that document is still in existence, by the production of a copy of that document or the material part of it, authenticated in such a manner as the Court may approve.

It is therefore important that a party using electronic information in the courtroom has a rigorous audit trail reliably logging when a drawing or document was created, every instance when it is sent or received, and if it has been amended (and if so, when and by whom). The court may need to understand how the original was turned into an electronic image stored in the system, then sent and received without alteration, up to and including its production in court. Arguments over admissibility of evidence can lead to investigations into the system that produced the paper, the method of storage, operation and access control, and even to the computer programs and source code. It may also be necessary to satisfy the court that the information is stored in a 'proper' manner.

Both Masons and Hammond Suddards Edge agree that the ability to show that an electronic information system is managed in accordance with internationally recognised and audited codes of practice or standards – such as BS7799 (ISO 17799) – will be persuasive to a court of law. BSI codes of practice[4] also provide guidance, and clients and their project teams should expect a software vendor and/or its hosting provider to have addressed the five principles of information management set out in PD0010:

- *Information management* Information types need to be identified and their different management requirements (e.g. security, format, retention, etc.) understood.
- *Duty of care* Demonstrating a responsible approach to legislative and regulatory requirements Evidenced, for example, by compliance with BS7799.
- *Documented procedures* Business processes and procedures (e.g. date/time stamping, version control, authentication, back-up, maintenance, etc.) need to be identified, specified and applied, and will cover both software and hardware issues. Again, compliance with best practice guidelines, for example, the UK government ITIL standards, would be a persuasive factor.
- *Enabling technologies* Component technologies (e.g. access control mechanisms, storage media, data integrity, compression techniques, etc.) need to be identified, specified and documented.
- *Audit trails* Can the system permanently and securely log details of each significant event in the life of a piece of information?

Part one of British Standard BS7799 (international standard ISO 17799 is based on the British Standard) lays down basic principles of information security management and sets out the code of practice to be followed by those managing electronic data. Part two, the Information Security Management System, covers certification processes to ensure compliance with part one and can have an important positive influence in determining the authenticity and accuracy of electronic documents that are submitted as evidence.[5]

7.2 Legal relationships with the software vendor

Entrusting mission-critical information created by a client's project team to a collaboration system places some strong obligations on the software vendor, particularly if it is functioning as an ASP. It is therefore vital that appropriate legal arrangements are put in place to support the key relationships:

- *between a software vendor/ASP and a client* (or its representative, for example, a construction manager); this may involve a Master Licence Agreement (MLA) (also called an ASP Agreement);

- *between a software vendor/ASP and end-users* (i.e. supply chain members) of the technology; typically, this involves an End User Licence Agreement (EULA) or some form of standard Terms and Conditions,[6] sometimes viewed on-screen when the application is first used.[7] Everyone using the collaboration tool should agree to use it on the same basis. Accordingly, there may be a strong argument for annexing the EULA to the appointments of the professional team and making it a term of those appointments that the consultants will enter into EULAs. This would also have the added benefit of highlighting the use of a collaboration tool at the earliest possible point in the project.[8]

Complementary information to these agreements may be contained in separate documents (forming part of the contract), for example

- *schedules outlining agreed levels of service delivery.* Sometimes described as the bedrock of contractual relationships with ASPs, SLAs specify how the service will be delivered and will focus on specifying the minimum performance levels of the hosting provider (whether that is in-house by a project team member, at the vendor's premises, or via a third party).

It may also be useful to specify relationships:

- between individual supply chain members stipulating use of the technology to communicate with each other. Making use of the system mandatory may require amendments to consultant agreements and/or contracts with the main contractor and sub-contractors (see earlier), or may be covered in specific project protocol documents (see Section 7.2.4).

7.2.1 Master Licence Agreement (MLA)

Describing the relationship between the software vendor and the ultimate client, a MLA should cover:

- grant of a non-transferable non-exclusive client licence to access and use the collaboration service in relation to its project(s)/business;
- restrictions on the use of the collaboration system (e.g. posting information that is illegal, defamatory, indecent, etc.; spreading software viruses; spamming, etc.);
- clarification of the parameters governing use of project data by authorised project participants only (including relevant staff, for example, from the software vendor's help desk);
- data ownership;
- terms whereby nominated participants will enter into an appropriate EULA with the software vendor;
- payment terms;
- copyright of the vendor's collaboration technology;
- software vendor's use of the client's branding and data;
- indemnification of the software vendor against misuse, unauthorised use, etc., of the collaboration system;

- confidentiality, including security and data protection precautions with respect to user names, passwords, personal information about users, etc.;
- termination provisions (see Section 7.3.2; also including what may happen to the data once the project is complete);
- jurisdiction (e.g. agreement written in accordance with English law; English courts to have non-exclusive jurisdiction);
- limitation of the software vendor's liabilities.

7.2.2 End User Licence Agreement (EULA)

An EULA will describe the contractual relationship between a project's participants and the software vendor/ASP. In many key respects, it will reflect conditions imposed in the MLA, but at a level specific to participants; for example:

- the participant's licence to access and use the collaboration system in relation to their role on particular project(s) and/or for particular client(s);
- terms whereby this licence might be terminated (e.g. if the client relationship with the participant is discontinued);
- grant of a licence to the software vendor to store and, subject to access privileges, to access and view project-related information where the participant owns the intellectual property rights.

Client contracts with designers usually provide that the designer retains the design copyright but grants a licence to the client and other team members to use the design in relation to the specific project. The EULA should reflect this principle, extending the licence to include viewing rights for the software vendor.

The MLA and/or the EULA will normally seek to limit the software vendor's liabilities for direct and indirect/consequential damages. Clients and/or end-users should check such agreements are well-drafted and that the vendor is not avoiding liability unreasonably – for example, it may omit liability for loss of data due to its own negligence, non-deliberate action, system non-availability, under-performance, or inaction, or it may limit liability to a small sum of money out of all proportion to the actual impact of any loss.[9]

Clients should also be careful to ascertain that a software vendor is adequately insured: can it meet its liabilities if sued by the client? Moreover, if the vendor is sued by another – perhaps larger – client (particularly if a court found that its liability limit was unreasonable), does the business carry sufficient cover to ensure that it can continue to deliver its services? If the vendor is a subsidiary of a larger group, should the client seek guarantees from the parent company?

Team members outside the contract between the software vendor and its client (or agent) may also need to protect their position by specific contractual conditions. Such second or third-tier team members may be instructed to use the vendor's services, but their client and the software vendor's client may be different organisations (e.g. the contractor and building owner respectively) and therefore the contractual links may be vague or non-existent.

So that these lower tier team members can protect their position regarding liability, they may, as a condition of joining the team and using the system, require specific amendment to their conditions of appointment. Typically, such amendments would seek to protect the team member against errors arising due to system malfunction, and ensure they have access to the data both during project delivery and for a suitable period after project completion. Absence of such conditions would seriously affect their PII cover. For example, the Wren Insurance Association Ltd (Whitton 2004) has suggested the following outline clause:

> The Client requires the Consultant and the other members of the Project Team to use a web-based information system provided by the Client ('the System'). The Client acknowledges that the Consultant shall have no liability for any costs, damages, expenses or other losses whatsoever resulting from breach of this Agreement and/or negligence on the part of the Consultant occasioned by the malfunction of the System. The Client shall ensure that the documents posted on the System can be accessed free of charge by the Consultant during the carrying out of the Project and for at least fifteen years from the date of practical completion of the Project.

It is important to examine the overall relationships on each project, which can be surprisingly complex, and develop a suitable solution to match. As with other rapidly developing areas, legal interpretations can change quite suddenly and need to be regularly reviewed.

7.2.3 Service level agreement (SLAs)

As described in Chapter 4, depending on the vendor, construction collaboration software can be hosted by a project team member or – on an ASP basis – by the software vendor or a third party hosting provider appointed by the vendor. Whichever route is selected, it is important to consider an SLA to underpin security and reliability requirements, etc. After all, even an in-house IT department will still be responsible for delivering a service to other team members, and an SLA will help to document the team's expectations and the department's own undertakings on performance issues – indeed, it can often introduce a degree of certainty to an in-house department's IT delivery capabilities that simply did not previously exist.[10] If hosting its own software, the vendor may specify an SLA direct with the customer. Alternatively, if the vendor outsources its hosting infrastructure, the specified service delivery levels will typically reflect the SLAs between the vendor and its hosting infrastructure provider.

The SLAs should include specifications relating to:

*   security provisions (including firewalls, intruder and virus protection, etc.);
*   provision of appropriate back-up systems;
*   compliance with the stated functionality;
*   ensuring the integrity of data processed and stored on the system;
*   creation of a full audit trail of the project (including storage of all current and prior versions of documents and information along with related comments, 'red-lines', 'mark-ups' and associated data, for example, 'action by' dates, 'reason for issue', 'instructions', etc.);

- provision of useful access to the full audit trail (e.g. a searchable database created and delivered from within the system);
- ensuring each user's identity (e.g. by entering a password and username – but, as with all such login, individual companies will need to ensure their employees exercise due care in safeguarding their individual access details; organisations should also manage security in relation to access by any temporary staff);
- provision of appropriate user access rights to access and view particular documents;
- sufficient levels of processor, system memory, disk space[11] and telecommunications bandwidth availability to allow adequate performance (e.g. response times when users interact with the software);
- levels of system availability (e.g. 99.5 per cent during working hours – but allowing for occasional planned (i.e. notified in advance) downtime for equipment or software upgrades);[12]
- provision for upgrades to latest software versions, and guarantee of continued compatibility of existing data with new functionality;
- clearly specified levels of customer support (e.g. helpdesk), perhaps defining response times, severity levels, etc.;
- extent and quality of end-user training.

What is covered in a SLA is open to negotiation between a client and the vendor/ASP, and different ASPs will, of course, have different capabilities. Within a project team, users may already have some experience of working with different ASPs and clients might consider user preferences when it comes to assessing SLA specifications. For example, an ASP might stipulate: 'the system is not available between 4:00 and 6:00 am daily, or as notified from time to time'.[13] Customers need to beware of imprecise definitions, particularly if it is unclear: (a) how much advance notification will be given, (b) how the notification will be delivered to the customer and/or end users, and (c) how long such periods of non-availability might typically last. Prolonged or frequent periods of non-availability can adversely affect efficient use of the collaboration system.[14]

7.2.4 Project protocol documents

Part of the standard implementation service offered by a vendor should cover the provision and tailored development of a Project Protocol Document setting out the standards or rules of operation for user companies working on the collaboration system. Typically, it will:

- provide common protocols describing how users publish, retrieve and manage information quickly and efficiently;
- establish security levels, including access rights for different team members and different types of information;
- be modified by the client and/or the project team to suit its processes;
- as a 'live' document, be maintained and updated as required by the client/team;
- need to be read in conjunction with non-project-specific guides on use of the collaboration system (e.g. user guides, etc.);

- detail the pragmatic working procedures to be followed by participants during any temporary suspension of service (such procedures are essential to ensure the integrity of the data once the service recommences – see Section 7.3.1).

Initially, such protocol documents tended to be very vendor- and project-specific, but in 2002 the growth of web-based construction collaboration technologies prompted the Building Centre Trust to start developing an industry-standard PIX Protocol (launched in March 2004). It covers a wide range of topics, from individual IT infrastructures, software applications, document and drawing systems, email distribution policies and internet usage policies to the standards and policies that need to be agreed and employed by the team as a whole (see also Chapter 8). Some topics have particular legal relevance. For example, the Protocol's 'Project leader checklist' (Building Centre Trust 2004; see Chapter 8) includes the following pointers:

- 'Ensure that the client implements the agreed PIX Protocol in all subsequent appointment contracts to the project.'
- 'If editable documents and data models are to be exchanged, agree the legal status for each format that is to be exchanged, determine the responsibilities of each party to the exchange and the verification procedure for the data being exchanged electronically.'…
- 'Agree security status for each document and agree who should have access to the document.'
- 'For electronic communications, consider adopting an Electronic Data Interchange Agreement.'
- 'Agree the status of email. In particular differentiate between "contractual" and "conversational" email.'

7.3 Service interruption or unforeseen termination

7.3.1 Managing service interruptions

As already mentioned, an ASP can take great steps to ensure its service infrastructure is as robust and resilient as possible (e.g. it can ensure high levels of redundancy: the availability of secondary, stand-by or back-up systems that can be made instantly available if a primary system develops a fault). But no technology is infallible. Birkby and Nugent (2002) suggest customers and users need to be clear about what happens if/when a project extranet system crashes:

> If the system crashes, there must be a back-up system that enables each of the participants – designer, contractor, subcontractor, client, to carry on working with minimal delay to design or progress on site. This may mean that whenever a document is uploaded onto the network, the person who has created or amended the document keeps a copy in their own electronic file, so it is available if needed. Everyone will, of course, revert to emails and faxes and circulation of hard-copy drawings until the system is back up and running, but there must be a procedure for updating the extranet as soon as it is back on line, so that everyone can have confidence in continuing to use it, and not resort to their own back-up system.

This is normally achieved by devising a series of rules for the operation of the extranet, sometimes called a protocol, which becomes one of the contract documents for each of the organisations working on the project.

A service interruption may just result in a delay, but, as we shall see, there have been (hopefully isolated) instances where electronic data has been lost. This risk, and the related issue of back-up (see Chapter 4), merits a little further consideration. For example, if the ASP (and/or its hosting provider) has an inadequate back-up regime (i.e. copies of all project data retained separately to allow later reinstatement in the event of a data loss), then a catastrophic failure may have serious consequences. Joch (2002) describes a situation that arose in an American project:

> Gensler, an international architectural planning and design firm…had hired a third-party service to host an extranet for sharing documents and e-mails with a client using a Web browser. The client originally asked that all communications, including requests for information (RFIs), be handled electronically and it didn't require anyone to make back-up hard copies. The project progressed as planned, a model of e-collaboration, until the extranet provider decided to upgrade its computer system one weekend.
>
> 'On the following Monday, we tried to look at some past RFIs,' recalls Bens Fisher, AIA, a Gensler vice president based in San Francisco. The firm discovered it was anything but a quiet weekend for the extranet's host. The upgrade had inadvertently wiped out all past RFIs associated with the project, destroying an archive of information that was essential for creating new responses. Fortunately, despite the client's 'electronics only' request, Gensler had made hard copies of everything…. That nod to tradition helped the architect reconstruct the lost information.

Again, the SLA between a project team and the ASP should stipulate what back-up regime is in place. For example, some ASPs back-up their systems every night so that, if there is a major failure, all project information up to that back-up can be reinstated, limiting lost work to only that undertaken in the period since the last snapshot of the data was taken.

Even if data is not lost, a service interruption can still be disruptive, requiring team members to change their working methods during the time the system is out of action. One solution might be to rely on email until the collaboration system is reinstated so that the integrity of the single central repository for all project data is not compromised. This needs to be clearly communicated to and understood by all project team members. BIW's project protocol, for example, stipulates the nomination of an 'information coordinator' who assumes additional responsibilities should the collaboration system stop working temporarily. To avoid confusion and duplication of effort, all emails and relevant attachments would be sent to the coordinator and it would be his/her task to input drawings, documents, comments, RFIs, instructions, etc. once the collaboration system was reinstated.

Although perhaps outside the strict scope of this book, it is worth mentioning that organisations need to develop and implement their own document retention policies: formal written guidelines about what type of information gets created, maintained, saved, archived or destroyed. Email is a particular issue here as many business-critical communications, ranging from contracts and other corporate documents to project-related documents and drawings, can be sent and received through this route, yet many organisations do not have clear policies on retaining and storing such emails.[15]

7.3.2 Managing ASP termination

Given the relative immaturity of the UK collaboration service market and the number of vendors competing for AEC market share, lawyers have understandably warned customers to take steps in case a chosen ASP may become insolvent. Birkby and Nugent (2002) suggest that the contract with the ASP should provide:

- a right to terminate a contract and transfer to an alternative service provider in the event of any doubts about the ASP's ability to continue providing services;
- an obligation on the ASP to provide assistance on transfer of the service to an alternative ASP, by providing access to all necessary records and data. Assistance in the form of consultancy services may also be required;
- an obligation to make records available in specified formats that can be accessed by the ASP's customer.

Birkby and Nugent urge prospective customers to monitor the financial status of their service providers so that they get early warning of any problems:

> insolvency of the ASP would be likely to cause serious problems and realistically, many provisions will become impossible to enforce in the actual event. So you may also wish to require the ASP to provide regular financial information, to try to detect early signs of any problems.

Of course, prevention is better than cure. As already stated, it would be prudent for clients contemplating a lengthy commitment to a business-critical relationship to seek detailed information about the financial status (e.g. audited accounts, management accounts, shareholder details, insurance cover, etc.) of their preferred provider(s) before entering into a contract. A vendor may ask the client to sign an appropriate non-disclosure agreement before releasing commercially sensitive details – understandable in a highly competitive market. However, clients will probably draw their own conclusions if a vendor appears reluctant to divulge detailed information about its financial position, either initially or at some later stage during the delivery of a project.

It would also be prudent to consider appropriate clauses to cover what happens should the software provider be acquired by another business. In the IT world, it is not uncommon for the new owners to offer reduced support for products and project services; some analysts suggest customers should retain the right to terminate the contract if their supplier is acquired, or stipulate that the acquiring company must take on the original contract's obligations on the same terms.

However, the demise of the selected vendor, while being disconcerting and disruptive, need not be a disaster. Experience in both Europe and America suggests that data *can* be transferred to a new provider with relatively little interruption to ongoing projects. When i-Scraper folded in early 2001, BIW quickly moved into the breach so that i-Scraper's UK customers and their project teams would not be inconvenienced, and BuildOnline did the same for i-Scraper's German customers. In August the same year, when US provider BuildNet filed for Chapter 11 bankruptcy, maintenance and support for its service continued uninterrupted, and during bankruptcy proceedings its customer contracts were among the first assets sold. But, as Berning and Flanagan (2003) admit, a mid-project move to an unfamiliar service provider does run the risk that the acquiring provider will not meet the expectations of the project team. Moreover, if the choice of a collaboration technology provider was a contentious issue between any of the project partners at the outset, the failure of that vendor can increase the risk of legal disputes.

Should transfer of data become necessary, clients should specify that any transferred records are delivered in an industry standard format (e.g. extensible markup language (XML); as mentioned in Chapter 3, the UK Network of Construction Collaboration Technology Providers is already devising interoperability standards to help ensure that information can be easily transferred to another system if, for example, a vendor went bust). Birkby and Nugent (2002) also urge prospective ASP clients to consider safeguards to ensure they can still access the collaboration system and the information it holds:

> If use of the extranet or the records it generates requires any proprietary software, then it will be necessary to obtain the source code for this (which most software providers will resist) or to provide for the code to be placed with a third-party escrow agent under an agreement that provides for it to be made available in the event that the software provider or ASP becomes insolvent.

Under a software escrow agreement, the software owner provides a copy of the source code for the software application to an agent (e.g. NCC Group – used by 4Projects and BIW – claims over 95 per cent of the UK market). Such agreements indicate the circumstances under which the code can be released to registered users. Typical 'trigger events' might include the bankruptcy or liquidation of the owner or its failure to maintain its software. Such breaches of a software licence or maintenance agreement would allow the registered user to invoke the escrow agreement. With access to the source code, the user could then provide it to an alternative hosting company, or run the software on their own servers, thus enabling business continuity. This type of insurance has already been used to rescue business-critical applications in the property sector.

Canary Wharf Group recently claimed release of two applications from escrow due to one of their software suppliers going into liquidation. Canary Wharf had invested a substantial amount of money and investment into developing bespoke IT systems to assist the organisation's construction and development plans. The systems provide the backbone to the group's development plans by handling the extensive paperwork process that is involved with any building project – from initial designs and planning permission, right through to site instructions. Without these applications being available through escrow, the cost, both in terms of lost time and development activity, would have had a major financial impact on their business.

Escrow provided Canary Wharf Group with the legal right and technical means to continue to maintain their software applications.

> While keeping a backup of the executable version of the software is always important, in the event that the application needs maintaining to keep it in line with a company's needs or the changing systems environment, only access to the source code will give the kind of protection that many companies require.
>
> (Leigh 2004)

Such contingency arrangements may also extend to ASP agreements with third parties (e.g. hosting partners) to provide continuity of service regardless of the ASP's status. For example, following an agreement with its managed hosting provider Attenda, if BIW ceased to be able to trade or continue operations, the BIW service would continue to be delivered for a minimum period of two months at no extra charge. This contingency arrangement is designed to allow sufficient time for former BIW clients to agree a longer-term arrangement with Attenda, to agree terms with an alternative hosting service provider, or – taking advantage of BIW's escrow arrangement – to run the service themselves on their own server(s). Sarcophagus offers a similar contingency arrangement whereby its third-party host maintains the-project.co.uk for three months allowing the customer to find an alternative provider without loss of service.

7.4   Discovery and the document audit trail

The transparency of many electronic construction collaboration systems can be a powerful stimulus to openness and the avoidance of disputes in construction projects. While it is perhaps utopian to believe that a collaboration system will completely eradicate all prospects of litigation, the use of such a system may at least reduce the time and costs associated with lawyers' document discovery processes.

In traditional paper-based construction projects, the project team may have exchanged literally thousands of pieces of information, including estimates, schedules, contracts, meeting minutes, change orders, RFIs, CAD drawings, submittals, maintenance manuals, correspondence, faxes, memos and emails. Some will now exist in paper form, some will still be on computers; some may be stored in archive boxes or filing cabinets, others may be stored on computer disks, CDs, removable hard-drives or tapes. The information may be spread across several different organisations involved in the delivery of the project, and, particularly in a multi-party dispute, several firms of lawyers may be simultaneously engaged in the discovery process. In such circumstances, the discovery of relevant documents can become an expensive, laborious, complex and time-consuming process.

However, when discovery is required for a project where construction collaboration technology has been employed, the cost, effort and time involved can be significantly reduced. All the drawings and other documents that have been published to the system are immediately retrievable, along with complete audit trails detailing who published them, when, and to whom they were issued, when they were opened, and by whom, and what comments may have been made relating to each document. The production of new revisions of those documents and any related status changes will also be detailed.

Of course, the completeness of this record will vary according to how extensively the team has used the technology. For example, where a project team has used email alongside the collaboration system, it will be necessary to instigate a separate discovery process to retrieve – if possible – relevant emails (and attachments) from each team member's email system.

7.5   Freedom of information

As well as facilitating access to information for legal discovery purposes, construction collaboration technologies may benefit some parties to a project, notably UK public sector organisations, by helping them quickly, efficiently and therefore cheaply meet their obligations under 'open government' legislation.

Local authorities in England and Wales, for example, have long had certain obligations under legislation[16] to grant the public right of access – subject to certain exemptions – to documents (agendas, reports and background papers) of the council, committees or subcommittees. The level of openness demanded has extended. From 1 January 2005, the Freedom of Information Act 2000 created a statutory right of access within 20 working days to all information held by upwards of 100,000 UK public authorities, though some exemptions still remain. Contractors, consultants and suppliers working for public authorities should be aware that information they provide could fall within the remit of the Act; details of their contracts, designs, etc. may be requested by, for example, members of the public, media organisations, pressure groups or competitors.

Where, say, a public body is perhaps the ultimate client or a partner in a PFI scheme, somebody might make a written request to the public body regarding information it holds relating to that project. Internally, the request might be forwarded to an official with collaboration system access rights, who can then use the system to view that information and, where appropriate and assuming it is not covered by any of the exemptions,[17] make copies for the purposes of satisfying the request. Simultaneously, that access and download would also be recorded, useful should the public body's compliance with the Act need to be audited.

7.6   Ownership of data and copyright

Sharing project information in an electronic collaboration environment can make some team members anxious about the ownership, use, and possible abuse, of the information they contribute.

According to guidance from the Construction IT Security Forum (CITSEC) (2004):

> In a collaborative environment, there must be specific provision made for the transfer of ownership of a data item as soon as it enters the collaborative environment. It would normally be expected that the client in a construction enterprise would be the ultimate owner of all the project data, although contractors sometimes take on this ownership and transfer it all to the client at hand-over.
>
> Ownership of the information also implies ownership of the environment. ... In the increasingly common case of the project extranet, there is currently still a degree of uncertainty among many participants as to the ultimate owner of the system – in particular when the system is hosted and provided by a third party. Logically, as part of the transaction which establishes the extranet in the first place, an explicit statement of ownership should also be made.

Such anxieties can also extend to intellectual property. For example, depending on what terms of appointment or project protocols have been agreed, architects and other designers may be required to submit CAD drawings in their native format (e.g. Autocad files in DWG format), giving rise to concerns about unauthorised use of design drawings – and hence breach of copyright – based on a perception that an electronic system may be more prone to abuse than a paper-based one. Subject to agreement within the project protocol, technical measures to reduce unauthorised use or alteration of documents might include issuing documents in 'static' formats such as PDF or as 'write protected' Microsoft Word documents, but the best legal protection is obtained by ensuring that contracts contain appropriate, effective and explicit confidentiality and licence provisions.

So far as internet-based construction collaboration technologies are concerned, there are no special rules relating to copyright. Protection of CAD drawings and other material is governed by ordinary copyright law (i.e. the provisions of the Copyright, Designs and Patents Act 1988 in the United Kingdom). As such, designers should follow normal procedures to protect their intellectual property. For example, RIBA recommends that architects:

- Include a disclaimer or statement of permitted use on all drawings. For example: 'This [plan/drawing] has been produced for [client] for the [project] and is submitted as part of planning application [application number/relating to site name] and is not intended for use by any other person or for any other purpose.'
- Include the architect's name and logo on all drawings and make sure that all work carries a copyright statement, for example, '© [name of copyright owner (UK)] [date of creation]'.
- Put a watermark through all drawings – this could be the architect's name or logo.

Client contracts with designers usually include explicit provisions about copyright in the designer's designs. Typically, the designer retains ownership of the copyright but grants a licence to the client and other team members to use the design in relation to the specific project (see Butler 2003), though there are some bespoke forms where copyright may be transferred to the client. The use of an electronic collaboration platform does not change such arrangements, though one can foresee instances where, say, a design may have been developed collaboratively – perhaps through commenting and red-lining – to such an extent that the final design no longer represents the intellectual output of one individual or firm but of a group, raising issues about co-ownership of the copyright in that design. Although instances of such collaborative design may initially be fairly rare, lawyers may feel that contracts should start pre-empting the issue by including provisions which cover this possibility.

Should a breach of copyright be suspected, any construction collaboration technology that maintains a full audit trail will at least be able to catalogue every instance of a CAD file being accessed, viewed or downloaded, detailing who instigated the action and at what date and time. Subject to the normal court rules (e.g. the Civil Evidence Act 1995), this audit trail can be offered as evidence. This contrasts with the situation with paper-based information, CDs or email attachments where it can be difficult, if not impossible, to show who has had access to a particular drawing once it has left the designer's office.

7.7 Archives

What happens to project data at the end of a project? Traditionally, team members would store a large proportion of the drawings and other paperwork in archive boxes and on microfilm, creating a considerable storage management challenge (a Construction Industry Computing Association seminar in November 2004 heard how, during 70 years of project work, multi-disciplinary consultancy Arup had accumulated more than 100,000 archive boxes, with an index held in 150 lever-arch files, and over 650,000 drawings held on microfilm).

Some construction collaboration technology providers seek to deliver their services for operation and maintenance or FM purposes beyond the end of the design and construction phases. In such circumstances, the project data remains online and can be updated as the facility in question is subject to repair, replacement of components, refurbishment, etc.; data is also maintained within the latest version of the software so that the customer does not need to worry about the continued availability of, and support for, older versions of the software, plus changing hardware technology, over time.

As described in Chapter 5, should the customer decide not to keep the project 'live', then the data could be stored in an off-line archive, incorporating all the drawings and documents held by the collaboration system along with the associated meta-data: database entries, issue lists, audit trails, comments, etc., and in-built version of the collaboration software to enable the customer to 'surf' the project data just as it was when it was live. BIW, for example, supplies its clients with an archive stored on a portable USB hard-drive. This can either be stored at the client's offices or downloaded onto the client's own network. Similar archives can also be created for individual members of the project team, covering all of the drawings, documents and meta-data to which they had access during the project.

Team members could also archive their project records to CD-ROM or DVD, but, particularly on large projects, this may mean production of more than one disk, and it may not be possible to replicate the functionality of the collaboration platform within the chosen medium. Moreover, as with many archiving issues, organisations need to consider the longevity of the chosen storage medium.[18]

7.8   Chapter summary

As this chapter has made clear, the emergence of construction collaboration technology has not significantly reduced the legal headaches familiar to many construction professionals. Indeed, it has added new areas requiring legal attention, particularly when it comes to documenting working relationships with the collaboration vendors and the effectiveness of the software and services they deliver. As a still-developing subject area, drawing on expertise from traditional construction law and from new areas such as IT, prospective customers and their fellow project team members should tread cautiously. They need to ask detailed questions of the technology providers, and to research their track records carefully. Most of the vendors have now accumulated considerable experience in supporting UK construction projects; there are some powerful case studies and numerous anecdotes – both UK-based and international – illustrating the key steps that might need to be taken; and some useful support tools are beginning to emerge (most notably in the United Kingdom: the PIX Protocol, and the NCCTP's data exchange standards). However, this remains a fast-changing legal area, and prospective users of the technology should always seek appropriate professional advice before making firm commitments regarding its use.

Chapter 8

Human aspects of collaboration technology

This chapter:

- refocuses the reader's attention on the importance of people and process issues in successful implementation of collaboration technologies;
- considers why individuals, departments and businesses might resist the idea of collaborative working;
- looks at the apparent practical barriers that might be raised to introduction and use of the technologies, including selection, timing, protocols, training and cost.

As discussed briefly at the end of Chapter 2, the movement of construction collaboration technology towards the AEC industry mainstream has met with some resistance. This is hardly surprising. Experience in other industries (e.g. IT, retail, manufacturing) suggests that failure to understand and adapt human behaviour, rather than technology, is the biggest single impediment to successful collaborative working. Arguably, in the more slow-changing – even change-resistant – AEC industry, culture, people or psychology issues are even more important:

> not a single project web site helps establishing ground rules, building team spirit, trust and commitment, takes in consideration the experience or seniority of individual team members, or accounts for the external social and business environment behind each team member.
>
> (Verheij and Augenbroe 2001, p. 3)

An industry rule of thumb suggests that successful implementation of collaboration systems depends 80 per cent on tackling the people and process issues and only 20 per cent on resolving the technology aspects. Accordingly, we can analyse the resistance to the technology by reference to two broad areas: resistance to the strategic principle of collaborative working, and resistance at a more tactical level to the adoption of the technology itself (distinguishing between the two may not always be easy: apparent concerns about the technology can often be used to disguise deeper issues relating to the whole idea of working in a more integrated, open and transparent manner).

8.1   Resistance to collaborative working

In the years since Sir Michael Latham highlighted the potential advantages to be gained through partnering approaches to project delivery (1994), industry estimates still suggest that the majority of the UK projects are still procured using more traditional, often adversarial approaches. This partly reflects inertia when it came to contemplating more open, collaborative approaches, whether this is found at the individual level, departmental or intra-organisational level, inter-organisational level, or within the industry as a whole.

8.1.1   Individual resistance

In AEC firms, individual advancement has frequently depended upon gaining time-consuming professional qualifications and years of project experience, using familiar, traditional, tried and trusted techniques. True partnering requires a more collaborative approach, and it can be very difficult to persuade individuals that they need to change, to adopt a different mindset and behave differently (*and* to use a new technology into the bargain), particularly if their entire careers to date have been devoted to achieving seniority through age and continual demonstration of their individual skills, expertise and experience.

For example, for many people within organisations, the predominant attitude to information has been to guard it carefully: 'knowledge is power' is a phrase often used. As a result, some individuals build entire philosophies about their roles and responsibilities based on a non-sharing concept. Collaborative approaches will be of little or no value unless people believe in them.[1] Once they accept the concept, they then need education, training and support to migrate from a non-collaborative mindset towards one in which collaboration is embraced both implicitly and explicitly.

Almost 10 years after his seminal report, Sir Michael Latham, in an entertaining column in *Building* magazine (2004), suggested there were six fundamental types of people who did not believe in partnering:

- The *stick-in-the-mud* who says: 'I've been with this company for 25 years. I've always got on okay, and I don't need a college boy telling me how to do my job.'
- The *jobsworth* well established, he says: 'I've been in this job for 25 years. ... Maybe I should go on a course and learn about [partnering]. The snag is that while I'm doing that, that whippersnapper who's just joined from university will get my job. So I'll stay put.'
- The *one who just doesn't get it*: 'Partnering? There's nothing new in it. I've always got on well with my clients, even if I only worked for them once.... I've never had any complaints from our subbies, and if I do I wouldn't give them any more work.'
- The *diehard sceptic*: 'I run this site, not the governor. When he comes, I'll tell him I'm partnering. Meanwhile, I'll carry on doing it my way.'
- The *control freak*: 'A good example is the designer who feels that the formation of an integrated project team marginalizes them. They would rather stick with

the traditional system of a disconnected series of performers with one contract administrator and leader – them.'

- Finally, the most serious case, the *young people fed a poisoned account of the way the industry should operate*: 'Take John and Jean, two students who graduate from the same university with a degree in quantity surveying. On day one, he is put next to a greybeard who tells him: "Look son, your job is to stop the greedy builder ripping off our council-tax payers." Jean joins a contractor as an in-house surveyor, and her greybeard says: "Look, luv, your job is to look for weaknesses in the works information to see where we can make claims and variations. We need to make the margin that wasn't in our tender." '

Attitudes to collaboration will also vary from profession to profession. Some workers will be well attuned to working in teams; others will see themselves as individuals. Verheij and Augenbroe (2001), for example, contrast the different approaches of contractors and designers, characterising the former as 'process workers' and the latter as 'knowledge workers'. For process workers, the team concept is simply part of their regular work environment, while in knowledge work, the concept of team is frequently associated with a loss of creative freedom and individuality:

> Engineers are trained to be independent workers... They prefer being measured on individual uniqueness and heroics not on collaboration and team behavior. ... The complexities and demands for speed of today's marketplaces are assumed to drive people towards increased collaboration and shared accountability. Paradoxically, they have the exact opposite effect on many knowledge workers.... The more complex a project becomes, the more an engineer wants to work in his or her own cubicle on a portion of the project, limiting any dependence on others.
>
> (Verheij and Augenbroe 2001, p. 17)

Of course, the distinction may not always be so clear-cut: there are some very team-oriented designers (and, doubtless, some individualistic contractors), but to varying degrees such 'creative isolation' can still be found among architects and other designers within project teams. This is perhaps not surprising. Traditionally, the planning, design and delivery of most projects tended to subdivided into many largely sequential processes and activities. Particularly in the early stages of projects, designers and other professionals can be inventive but somewhat secretive about their creative ideas, only exposing them to critical review (e.g. by constructors, materials suppliers and/or component manufacturers) when they are quite advanced.

Moreover, the sometimes confrontational or adversarial nature of traditional UK projects can preclude genuine teamwork, as participants seek to avoid risk or blame should things go wrong. As Latham's stereotypes (2004) make clear, older designers who may always have worked in such a way might seek to justify why they should not change, and will need active encouragement and support to make the necessary changes. At the same time, there is an opportunity for those teaching younger or new industry professionals (educational institutions, professional bodies, mentors, etc.) to begin to promote more collaborative attitudes and approaches. As Latham sums up: 'It needs serious training, deep culture change led from the top and continuous reinforcement.'

8.1.2   Intra-organisational resistance

As Latham suggests, even if individuals do move towards more collaborative approaches to their activities, this may count for little if their employers do not also encourage and support such approaches. Just as individuals have often adopted an attitude of 'knowledge is power', within many organisations, there can be departmental resistance to the notion that they should share information. Key functions – sales, IT, procurement, HR, accounts, etc. – often sit in 'silos', with their own agendas, systems, attitudes (including the often destructive 'not invented here' syndrome), and varying degrees of influence over corporate strategy. There may also be regional silos within which different parts of an organisation pursue regional agendas that differ from each other, and from that of head office. Most construction businesses have yet to resolve this challenge, let alone the related challenge of creating an environment that encourages collaboration.[2] However, if a business can begin to identify where, why and how collaboration might deliver business benefits, it can begin to justify making some quite profound changes.

Organisations may need to alter their organisational structures and cultures, to change their internal management processes and to promote a different style of leadership if their staff members are to succeed at working in teams. In this respect, it is encouraging that some major UK construction businesses have already begun to preach collaboration and openness as part of their corporate creed. Carillion, for example, states that its core values are: openness, mutual dependency and collaboration.

People and process adaptations are a vital precursor to the successful integration of IT into an organisation's operations. For example, within the UK construction industry, the Building Centre Trust (Goodwin 2001) stated: 'It is our belief that the technical barriers are far less significant than the organisational and managerial issues which have to be addressed' (p. 7). Some management writers even discount technology altogether; US economist Paul Strassman, for example, argues that there is no link between technology investment and productivity – in his view, it is all down to good management, and good managers are better able to exploit technology. In short, organisations need to understand and develop the collective experience, knowledge and wisdom of their own people, and then to devise and implement supporting cultures, structures, systems and technologies.

For example, managers could amend employee job descriptions to emphasise team performance and, while accepting there is still room for individual brilliance, place less emphasis on individual achievement alone. Senior managers ought to be seen to preach *and* practice collaborative working (sometimes described as 'talking the talk, and walking the walk' – as distinct from those who are 'talking not walking'). Collaborative working should be rewarded, thus motivating and incentivising employees to change their attitudes and behaviours. The objective, according to Davis (2003, p. 22) is to: 'to create an environment where people are not only comfortable, but also positively enthusiastic about collaboration'. The incentives may be financial, although pay may not be the key improvement demanded by employees: 'Intangibles such as a feeling of involvement in a task, acknowledgement and celebration of success, and, most importantly, coaching someone else in the job, can help drive the adoption of collaborative technologies'.

Business process legacy can also inhibit effective collaboration. Organisations may be tempted simply to carry on doing things the way they always did – ignoring the danger that, by doing so, they will always get what they always got. Instead, they may need to challenge accepted processes and thinking. Where collaboration tools are applied in organisations that do not encourage collaborative processes, it is not surprising that the tools may appear to fail.

In short, to avoid accusations that they are only paying lip service to the principles of collaborative working, organisations must resolve any internal issues they have about collaboration *before* they start considering committed, collaborative relationships or alliances with external clients, partners or suppliers. Once these intra-organisation issues are eradicated, focus can switch to breaking down the inter-organisational fear and mistrust that often exists.

8.1.3   Inter-organisational resistance

Traditionally, project participants in the UK AEC industry have established external trading relationships based on short-term commercial outcomes relating only to the immediate project. Essentially, the approach was adversarial, focused on cutting costs/maximising profits from the transaction, while minimising defects and delivering the project on time, with onerous contracts setting out participants' roles and responsibilities and managing their risks.

But, as outlined in the first part of Chapter 2, collaborative working has begun to drive change. Hierarchies are eroded and, instead of the traditional handover of responsibilities (e.g. from architect to contractor to facilities manager), all parties take a keen interest in the whole process of delivering the project. There is also increasing emphasis on longer-term relationships, helping partners to focus on working together and delivering mutually acceptable levels of profit over a series of projects, and on achieving and sustaining improvements in design, service quality, health and safety performance, innovation, etc. The desired benefits also include more qualitative 'relationship' factors such as improved co-operation, fewer disputes or claims, better communication, etc.

When working in a more collaborative environment, the relationships between organisations become particularly important. The internal culture of each prospective partner becomes a good starting point. Advocates of strategic partnering relationships understand this well: partners should: 'analyse each others' goals, philosophies and cultures' say Bennett and Jayes (1995, p. 50); staff on both sides in a potential long-term relationship may need to display 'alliance competence' (Spekman and Isabella 2000), so that they can begin to understand and align their expectations and responsibilities accordingly. In many instances, the early stages of a partnering arrangement often involved facilitated team building and workshops focused on breaking down traditional hierarchies, on communication, listening and win–win approaches to problem solving.

While collaboration may require organisations to adapt their internal structures, cultures and managerial processes, the change may also extend to how they encourage and support – instead of block or stifle – collaborative processes across organisation boundaries. For example, organisations might consider how employees could be encouraged to work as staff of a 'virtual company', perhaps incentivised on the extent to which they contribute to the mutual objectives and success of that entity.

This may mean, in many situations, changing how businesses collectively handle particular processes, as Zara Lamont, chief executive of the Confederation of

Construction Clients, pointed out in 2002:

> What [clients] really want to see is IT adding value to their own operations. ...If they [business processes] are the wrong processes, it will mean we are doing the wrong thing more efficiently. As a cash-rich, time-poor society, what really makes the difference is communication and collaboration.... We don't need e-commerce to do this. We need teams, chains and clusters who are prepared to work together over time in an open and honest way to develop the right processes.
>
> It will take considerable investment by all parties to sort out the right business processes, software interfaces and protocols. Having done that, you will then need to review and improve them in operation....
>
> The more you put into getting the processes right and changing the culture, the quicker the payback. But you don't want to be climbing the same learning curve on every project – hence the need to work as integrated teams over a series of projects.

8.1.4   Industry resistance

Many of the issues raised in Sections 3.1.1–3.1.3 contribute to an overall industry resistance to the notion of collaborative working. The contrast between the AEC sector and other industries was underlined by the comparison of the construction sector by Green *et al.* (2004) with the aerospace industry (Table 8.1), which highlighted how the structure of the former created an underlying low climate of trust.

Echoing a point already made several times, much of the UK AEC industry is 'heavily shackled by a conservative culture and mindset [and] inhibited, in many cases, by organisational inertia' (Autodesk 2003, p. 1); it is very fragmented, with the danger that collaborative working becomes the norm for a progressive minority, while the majority of the sector continues in its old ways; and most project teams tend to be focused on design and construction, not on the 'whole life' of a built asset. While there have been some clear examples of the benefits of partnering (e.g. Crane and Saxon (2003, p. 56) mention several successful strategic arrangements: Taylor Woodrow with both Tesco and Shell, the BP/Bovis Alliance, Mace with Cannon Healthcare, and HBG with Argent Properties), the approach has yet to become the accepted norm across the UK construction industry. It may require, as Green *et al.* (2004)

Table 8.1 Structural comparison of aerospace and AEC industries

Aerospace – high trust economy	AEC – low trust economy
Highly consolidated	Highly fragmented
Few customers	Many customers
High knowledge intensity	Low knowledge intensity
High barriers to entry	Low barriers to entry
Long time frames	Short time frames
Fixed locations	Transient locations
High inter-dependency	Low inter-dependency

Source: Adapted from box 4.2, Green *et al.* 2004, p. 29.

argue, that progressive minority to demonstrate sustained business advantages before more of the rest begin to change their own cultures, processes and styles of service delivery:

> The emergence of integrated procurement approaches in construction such as prime contracting and PFI is causing a polarisation in the construction market. Firms have strategically positioned themselves to take advantage of new markets. The competitive advantage of the leading players will increasingly be based on their skills of integration and supply chain management. These emerging niche markets already present significant barriers to new entrants.... Clients may benefit through a more integrated service. Integrated supply chains potentially stand to benefit by competing primarily on the basis of innovation and expertise rather than cost.
>
> (p. 36)

The partnering atmosphere of mutual trust, openness and co-operation is, of course, promoted through use of construction collaboration technology, and many of the early technology case studies tended to feature projects where the partnering ethos was strong. In such a transparent environment, everyone knows who did what and when; it was easy to see who was responsible for a delay or problem. This will, however, have done little to persuade more traditionally minded individuals and businesses who had yet to commit, or even had no intention of committing themselves to partnering approaches. Early research evidence suggested there was still some way to go:

> certain sectors of the construction industry showed greater resistance than others to the implementation of construction project extranets...Regarding contractors, [there] seems to be a significant divide between those entrenched in more dated methods of operating and those who have adopted a more progressive approach. In this case progressive refers to companies that are actively integrating partnering, supply chain management and other such core developments into their management process.
>
> (Murphy 2001, p. 67)

8.2   Resistance to collaboration technologies

To recap the previous section briefly, it will, of course, take a long time for many individual industry professionals to adopt more collaborative attitudes and behaviours, just as it will take many construction businesses a long time to alter their organisational structures, cultures, management processes and leadership styles. Equally, these businesses will need to identify and then work with other like-minded organisations before they can jointly develop new, more appropriate inter-organisational collaborative processes. And it will take even longer for the early collaborative alliances to demonstrate sustained business advantages and convince the rest of the industry of the strategic need to adopt more progressive attitudes and behaviours.

In the meantime, however, we can now consider the construction industry's resistance to technology, and to collaboration technologies in particular.

8.2.1   Is the UK construction industry techno-phobic?

As mentioned in Chapter 2, the industry has been keen to use new IT tools where appropriate, and while it may have lagged behind other sectors in adopting some technologies it would be inaccurate to describe it as completely techno-phobic.

As with most industry sectors, attitudes to IT vary. For example, Autodesk's 2003 industry survey suggested there was a division on future strategy between 'long-termists' who see quality, collaboration and effective partnerships as primary drivers and 'short-termists' who focus on cost and immediate return, with the latter more common. While most respondents recognised the need for change (e.g. partnering and integration), they were unsure about implementation, doubtful about who stood to benefit most, and – apart from IT – negative about all the supposed facilitating factors, especially culture and mindset: 'the industry's heritage and innate conservatism is, they confirm, a source of major weakness' (p. 5). Looking more specifically at extranets, Croser (2003, p. 73) suggested there was 'fear, uncertainty and doubt (FUD)...somewhere between apathy and hesitation'.

These cultural barriers to technology adoption may also relate to the age of many decision-makers, as Becerik (2004b) identified:

> On the company/customer level there is often an age barrier, which undermines technology adoption initiatives in many industries, not just AEC. Owners and top management of the larger, industry-leading AEC firms are generally over 45 years old. Although there are a growing number of technology advocates in that group, there is still a large percentage who are either reluctant users of technology in their work, or who have simply avoided dealing with it....These are not likely adopters of web-based technologies, which require both frequent presence online and comfort with invisible transfers of large amounts of their companies' valuable information. They will resist change unless they see a clear value and are either personally comfortable with, or directed by a client to make the investment.
>
> (p. 241)

Attitudes to technology can also vary according to the size of the AEC organisations involved.[3] Looking at contractors, for example, Stratagem/DTI (2003) found project collaboration was an important e-business issue for 57 per cent of companies with more than 100 employees but was important to only 35 per cent of companies with between 50 and 100 employees. They noted: 'The gap between larger companies who want to use project collaboration with their supply chain is limited by the inability of smaller sub-contractors to take it up' (p. 27). Their study highlighted: 'a large gap between large main contractors activities in e-business, (91 per cent use email, 58 per cent project collaboration), with their supply chain and small sub-contractors ability to respond, (only 48 per cent use e-mail and 12 per cent project collaboration)' (p. 28).

8.2.2   Moving from 'novel' to 'normal'

Naturally, one might also expect attitudes to technology to change over time. As particular technologies become more accepted and more commonplace – as they make the transition from 'novel' to 'normal' – their take-up by construction industry organisations will also increase (look at the take-up of fax machines and computers as examples). As far as adoption and use of web-based construction collaboration technology is concerned, this transition will partly involve learning more about the kind of technology issues that have been covered in the preceding chapters, namely:

- *Dot.com doubt* (see Chapter 3) There were too many competing systems. Most of the providers were relatively new and had little track record. There were concerns about who would survive, and which system (if any) might become the de facto standard, and the ASP model of software provision was still relatively unfamiliar.
- *Hosting hesitation* (see Chapter 4) Many AEC organisations had traditionally managed their own IT systems; the ASP was sometimes seen as a threat to in-house IT departments, and some 'dot.com doubt' remained over hosting. Without direct control over the systems, how could team members be sure about system performance, availability, reliability, security, support and scalability?
- *Feature fears* (see Chapter 5) It was difficult to choose between the different systems, particularly when most of them looked fairly similar and all claimed to offer essentially the same features. At the system organisation level, some users have been particularly concerned about controlling confidentiality and trans-parency.[4] At the communication level, the key issues have been sharing drawings, enabling effective feedback, and ensuring a clear audit trail. At the management level, teams wanted to be able to manage common project processes such as transmittals, RFIs, etc., according to their own approaches. And many team members have been anxious about how easy to use the systems might be, and about what level of vendor support they will get during implementation and training and throughout the project or programme itself.
- *Connectivity concerns* (see Chapter 6) For many would-be users, the advent of internet-based collaboration solutions posed questions about their existing software, hardware and telecommunications systems.
- *Legal logic* (see Chapter 7) In a risk-averse industry hitherto renowned for its often adversarial attitudes and heavy reliance on contracts and other legal reme-dies, there were bound to be concerns about the legal implications of using elec-tronic as opposed to paper-based communications. Software providers now become part of the project fabric, and team members need new agreements regard-ing the provider's role and responsibilities, and issues such as data ownership, copyright and access to data in case of disputes.

Generally, resistance focused on any or all of these issues can be addressed by more detailed consideration of each topic (this book will hopefully have helped, and should also point the reader towards other information sources such as industry organisa-tions and publications), by talking to existing customers, end-users and the vendors of the various different systems, and by taking advice from in-house and external experts (e.g. IT staff or consultants, lawyers, etc.).

8.3 Managing the human/technology issues

However, several further and sometimes inter-related practical issues remain if we are to support the effective selection, introduction and implementation of construction collaboration technologies by the project or programme team. Some of these were confirmed by respondents to two US studies, by Becerik (2004a) and Nitithamyong and Skibniewski (2004). Becerik's suggested 'critical enhancement areas' included several technology improvements (e.g. intelligent workflows, improved connectivity, integration with fax, email and other desktop applications, greater interoperability), but also highlighted that the most important problems are still organisational: 'The organizational issues surrounding the use (who is in control) as well as the psychology involved in getting all participants in projects to accept using new technology are now in focus' (p. 11).

What kind of organisational issues might be involved? Nitithamyong and Skibniewski (2004, p. 38) identified 11 variables that they judged to be critical to the performance of a collaboration application on a construction project. Excluding the factors already outlined (support levels, functionality, security, reliability and levels of integration were dealt with in Chapter 4; internet access availability and type of internet connection were dealt with in Chapter 6), the organisational factors included project type and duration, usage frequency of advanced features, ability of project manager(s) and – most critically – levels of internal support. The latter included: an ability to align implementation strategy with the project team's strategy, user participation prior to participation, top management support, training provided, and availability of resources (money, time and personnel) (p. 31).

Similar technology adoption issues were also highlighted by a 2004 ITCF survey (see Tables 8.2 and 8.3) and by the DTI's 2004 benchmarking survey (Table 8.4).

The remainder of this chapter will focus on many of the issues identified by the US researchers and by the ITCF survey respondents, namely:

- *Selection* Depending on what type of owner and what type of project is involved, how should we choose a system?
- *Timing* When should we introduce a system?
- *Encouraging buy-in or take-up* How do we promote 'internal support' and overcome 'FUD' (Croser 2003), conservatism, scepticism or poor initial experiences?
- *Agreeing exchange standards* How do we ensure everyone adopts the same processes, naming conventions, file formats, etc.?
- *Training* Do users require extensive training? What tasks are changed or replaced? What new responsibilities arise?
- *Cost* Who should pay for the system, and how should the cost be negotiated? Is there a minimum size of projects below which collaboration technologies will not be feasible? What additional costs might be incurred?
- *Differentiation, or competitive advantage* How will it give us an edge over the competition or make us stand out from the crowd?

8.3.1 Selection

Obviously, how an organisation goes about choosing a collaboration technology will depend upon its role and upon its project or programme needs. For example, many if not most AEC industry clients only procure a new building or other asset very occasionally. They may, therefore, want to be advised by industry professionals (e.g. an

Table 8.2 Barriers to technology investment

How significant are the following as potential barriers to your investment in IT?	Per cent of respondents
Technology is too expensive	57
Technology changes too quickly	50
Inadequate impartial advice available on the best way forward	49
New technology needs new ways of working	48
Technology is unproven	45
Associated costs are too high	45

Source: Adapted from ITCF 2004.

Table 8.3 Boosting technology investment

What would help your company in terms of encouraging greater use of IT?	Per cent of respondents
Reduction in cost of IT	60
More user-friendly systems	55
More training	55
Improved awareness of benefits	51
Improved software compatability	50
Setting industry standards	46
Improved capacity for emailing large files	35
Improved site access to IT	34
Improved capability to manipulate drawings	32
More portable computers	30

Source: Adapted from ITCF 2004.

Table 8.4 Obstacles to technology implementation

Can you tell me what has made it difficult for you to implement technology?	Per cent of respondents
Set-up costs	33
Running costs	19
Lack of skills (staff)	14
Lack of time/resources	12
Lack of knowledge	8
Reluctance of staff	8
Difficulty integrating IT systems	7

Source: Adapted from DTI 2004, p. 39.

architect or engineer) as to which, if any, IT solution would best suit delivery of their scheme and their future information needs.

On the other hand, a much smaller number of industry clients procure new buildings or other assets much more frequently. Central and local government departments and agencies, utility businesses and transport organisations, for example, are constantly replacing and updating key service delivery assets;[5] major retailers and financial institutions may have ongoing programmes of simultaneous new-built, refurbishment and extension projects; and property developers may be engaged in a steady series of new-built schemes. Such prolific organisations may have dedicated in-house departments, they may have long-term partnering framework deals with particular AEC professionals, contractors and suppliers, or they may appoint teams on a project-by-project basis.

Increasingly, however, regardless of how projects are delivered, the growing emphasis on collaborative approaches has begun to influence the more prolific client organisations (and their partners), to look at adopting a single construction collaboration solution that can be employed across all their schemes (similarly, new focus on 'whole life cost' has led some clients to look at solutions that can be used to manage their completed facilities after construction has finished). Up to the mid-2000s, only a few major UK clients had made such decisions (some were mentioned in Chapter 2). Many consultants and contractors, while sometimes favouring particular systems, will still tend to follow any stipulations made by a client, even if it means staff need to be proficient in the use of several systems at once.

Selection processes will vary from client to client and/or from project team to project team (but should cover all of the areas outlined in Section 8.2). As with many other aspects of the AEC industry and of business and commerce generally, the tendency would be for the client or project team to invite proposals from more than one provider. Some buyers initiate the process by sending out detailed questionnaires (similar to those used in industry vendor surveys, e.g. CICA 2003), then shortlist a few providers to present more detailed proposals and to demonstrate their applications to the team; others might skip the first stage and perhaps rely on desk research or word-of-mouth recommendations to draw up a short-list. To help encourage later buy-in and take-up of the selected system, it is common to involve as many members of the project team in the selection process as possible. Assuming, say, 2–3 vendors pass this stage, industry references may be taken up, the CVs of implementation, training and support consultants might be vetted, and a 'due diligence' process may also be undertaken to check, say, an ASP's hosting facilities, its financial stability and the legal protections built into contracts and SLAs, etc. Given that most of the UK construction industry remains very cost-conscious, price will often also be a key factor (see Section 8.3.6).

To date, rather than adopt a 'big bang' approach, most potential customers have adopted a cautious approach to rolling-out the technology. Typically, they might decide to test a system on a pilot project before making a decision about standardising on it for further schemes. If the first pilots do not go well or are inconclusive, then they may try similar pilot exercises with other vendors' products.[6]

The type of project can also be a factor (Nitithamyong and Skibniewski 2004). Simple residential projects, with fewer team members and less communication barriers, will often see little benefit from using a collaboration application, but more complex schemes – for example commercial buildings, heavy civil engineering and industrial projects – tend to be more expensive, more time-consuming, have bigger teams, and usually generate higher volumes of drawings and other documents, and will thus benefit more from having an application to manage such information.

8.3.2   Timing

Just as with any change initiative, industry experience to date (with research confirmation by Nitithamyong and Skibniewski) suggests that the best time to introduce collaboration technology to a construction project team is as early as possible. Where a scheme is still in the very early stages and some doubt remains about whether the project or programme will go ahead, then it may not be appropriate to introduce the technology (unless, perhaps, the vendor is prepared to share that risk – as happens on some PFI/PPP bids). However, if detailed design is already significantly advanced or if construction on site has already commenced, then it may be too late to achieve any real benefits; indeed, team members may even demand additional fees as the system's introduction may have required additional project inputs. In between these two extremes, the optimum time to introduce a collaboration system would be the point when team members are appointed and fees are agreed to move the scheme forward from the conceptual stage towards a more detailed consideration of design and buildability issues, that is, when the initial project team begins to expand significantly and include individuals from other disciplines and/or companies with whom the core team need to collaborate. Working with the vendor's consultants, team members can then get involved in the 'nuts and bolts' of how the technology will be configured to suit the project and their inputs to it.

There is also a timing dimension to how the project team begins to work together. Teams are sometimes said to go through four stages during group formation – forming, storming, norming and performing (Tuckman 1965) – and the interpersonal skills of a vendor's staff can play a critical role in promoting the collaborative effort required at an early stage. Even after initial selection meetings (forming), the vendor's consultants may still have vigorous debates with team members about the best ways to use the system (storming) before they can finalise the detailed structures and processes in project protocols (norming), provide face-to-face training, and – post-implementation – continue to support the project team's evolving demands of the system (performing).

The longer a project has continued before such technology is introduced, the bigger and more time-consuming the 'backfilling' task will be to populate the system with pre-existing project data (assuming the client or team wants this, of course – many may simply start with the latest or most current set of drawings and other data). Moreover, when initial team member staff are already routinely committing their data to the system and using it to manage project processes, it becomes easier to convince new joiners to regard the system as 'the way things are done' and adopt the same attitudes and project practices.

Time is, of course, a finite project resource and the range of features offered by a collaboration system can have an impact on how quickly it can be introduced into the project team. For example, accepting a pre-configured workflow shortens the set-up time, but if it requires team members to change their existing working methods they may need some training. On the other hand, more flexible systems allow users to set up workflows to accurately match how they prefer to work; this reduces the learning curve but increases the set-up time (however, as the workflows might then be duplicated on future or parallel projects, the additional set-up time may only be an issue for the first implementation).

One final time aspect relates to the ITCF survey (2004) respondents' view that 'technology changes too quickly'. Waiting for the pace of technology change to slow down may have been advisable when collaboration systems first hit the UK market, particularly as a few were US-based and needed adaptation to suit UK construction processes, but by the mid-2000s the leading vendors' systems were no longer changing significantly from year to year and the core features were broadly similar. Clearly, it would be unrealistic for any technology vendor to stop developing its system altogether solely to attract change-averse customers, but, in this respect at least, the ASP model of software delivery (see Chapter 4) does allow an application to be developed incrementally with small improvements and new features being added and made available gradually so that the impact on end-users is not excessive.

8.3.3   Encouraging buy-in or take-up

As mentioned already, involving end-users in the selection process is a key stage in achieving buy-in to the adoption and use of a construction collaboration system. If a team has been, or is being, brought together to work on a partnering basis, one might expect buy-in to be achieved quite quickly. Conversely, any vestiges of old-style adversarial approaches may mean that some team members harbour FUD about the technologies and how they might be employed – they may feel uncomfortable about the motives, perhaps viewing it as a 'Big Brother' exercise.

It is also worth examining people's different approaches to change. Ignoring for a moment the shift on a minority of projects from traditional project delivery to partnering, the switch from conventional distributed and largely paper-based information systems to ones employing shared centralised repositories of project data can also seem a major step. This may be a 'generational' or 'Luddite' attitude, with older and less IT-literate professionals resisting new methods, while their younger, more flexible and more IT-literate counterparts embrace it (as one architect told the author: 'many younger professionals start university already using computers and utilise the tools continuously so that it becomes second nature. You even get students coming through university these days having never used a drawing board'). Alternatively, it may simply reflect a high degree of conservative inertia. Change certainly needs to be carefully managed as – apart from people who regard themselves as innovators – most people tend to be, at best, passive about, and, at worst, resistant to change:

> Constituting about 10 percent of the population, innovators intuitively see the possibilities of a new technology and want to use it to generate change in their organizations. At the other end of the curve are the 10 percent who are skeptics and who will resist any changes on principle. The vast majority of the population are pragmatists, people who are of the 'show me' type. That is, they need to see proof that the tool will have a positive effect on the bottom line.
>
> (O'Brien 2000, p. 35)

Accordingly, when the new technologies were first introduced they needed to be sold first to the innovators and then given time to mature and become acceptable to the pragmatists. O'Brien (2000) says this can pose problems for a construction collaboration system (or 'project Web site', as he calls it):

> The problem with a project Web site, which is meant to draw the project team together, is that innovators and pragmatists must use the technology simultaneously. (Generally, skeptics will find a way to avoid using the tool). This creates a barrier at the site because the pragmatists are not easily sold on the use of an immature technology. Successful implementation requires extensive championing on the part of the innovators.
>
> (p. 35)

The importance of champions is recognised by most vendors, who will tend to suggest that individuals are appointed to become the project team's 'hub' for communications. The value of having a person in such a role was also stressed by research results from the United States, and by the UK's Construct IT:

> the presence of a champion along with the power of the owner to mandate is the most important factor contributing to the successful implementation and usage of [extranets]. The champions have to be technology savvy, understand the necessity to use this tool, the technology, and what it is offering for the job. Champions should be proactive and connected to the team. Success starts from the top and there should be a top-down buy-in by the project team.
>
> (Becerik 2004a, p. 9)

> it is recommended that the use of a [collaboration application] in a project should not be solely decided by an owner or the system will tend to lack a champion and could potentially result in failure. Project teams should identify a champion (or even assign a member to act as a champion) who could encourage team members to use [the application] prior to its actual implementation in order to ensure its success.
>
> (Nitithamyong and Skibniewski 2004, pp. 18–19)

> Once a project sponsor has been clearly identified, it is as important to identify a champion to manage the e-Project information. This champion can either be an independent specialised person or consultant not otherwise directly involved with the design, planning or costing of the project. Alternatively it could be the lead consultant such as the project manager or architect. In either case, it is important to check that the Project Sponsor, to effectively carry out their role, has entrusted the appointed Project Information Manager with the sufficient powers.
>
> (Construct IT 2003, p. 7)

As described in Chapter 2, construction collaboration technologies were first successfully deployed in the United Kingdom on projects that tended to have a strong partnering ethos, and so project team members were more likely to be receptive to the change and to the introduction of new technologies. Many of the 'show me' pragmatists on these early schemes gradually recognised the positive effects of the technologies, and – particularly if they have employed them on a succession of schemes – have effectively become innovators or champions themselves and helped to encourage new groups of pragmatists to use them.

Where such 'viral marketing' effects are not possible – on new projects undertaken by teams of so-called 'extranet virgins', perhaps predominantly drawn from SMEs, for example – the system providers' implementation and training staff will tend to focus on educating and managing the expectations of both enthusiasts and pragmatists about the technologies. As with many change initiatives, it is vital to help individuals identify: 'what's in it for me?' This may be a simple process, perhaps contrasting familiar frustrations with existing methods of project communication (e.g. working on out-dated designs, time wasted in searching for documents, slow or incomplete communications, etc.) with the individual benefits to be gained through the brave new world of electronic collaboration (e.g. no more repetitive form-filling, task simplification, etc.).

There will also be advantages at a company level too. For example, in an analysis undertaken by Stratagem/DTI (2003), 'SME inability to electronically transmit and receive CAD and other information' was regarded as a significant industry weakness:

> The inability to transmit and receiving CAD files limits the SME to receiving and sending drawings via the post. The ability to work with CAD electronically creates an opportunity to speed up the processes associated with winning business and dealing with any client drawing changes. Most respondents saw this step as crucial in the development of an effective basis from which to engage with large clients.
>
> (p. 35)

Just as the development of partnering relationships often involves workshops, team-building and other communication exercises, so the introduction of collaboration technologies needs to be assisted by disseminating information, running workshops, by agreeing working practices, by training, and by providing test-bed environments in which new users can experiment with the new technology, feel their way forward at their own pace and discover the benefits for themselves.

> The more team members, or end users of the system, who became involved during the planning process of…implementation, the more likely it was that the system would yield higher performance. User participation during the planning process also had a significant positive correlation with team members' attitudes toward [the technology]…The common sense reasoning is that if team members have participated in the development and implementation process, then they will be more likely to develop a better understanding of how the system can assist them in performing their jobs effectively.
>
> (Nitithamyong and Skibniewski 2004, p. 19)

If scepticism cannot be tackled by 'carrot' benefits, then it can, to some extent, be enforced by a 'stick' approach, perhaps making use of the collaboration technology mandatory for all project-related communications through contracts and project protocol documents at the inter-company level, along with individual contracts and job descriptions at the user level. This can help avoid the problem whereby some project participants choose to communicate via parallel systems (fax, post, courier or email, for example), bypassing the collaboration platform, creating information gaps in the central repository in order to keep non-users in the loop, and duplicating effort and data. Magub and Kajewski (2003) noted that an internet-based system might be regarded by some participants as multiplying the number of communications, and ranked this perception as the second highest barrier to effective project participation (behind opposition on the aforementioned 'short-termist' grounds of cost and technology).

Of course, it will help if, the first time participants use the system, they can locate the information they need. Collaboration tools cannot be implemented in a vacuum, they need to be populated with information around which people can then collaborate; O'Brien (2000, pp. 38–39) advocates 'seeding' the system with useful information to give project members a good first impression. Most providers will support such work, their consultants helping to 'backfill' the system with key documents during implementation and commissioning. Once it becomes clear that the collaboration system is the best source of up-to-date, accurate information, scepticism may quickly be replaced by a more pragmatic attitude. Such attitudes will also be reinforced once it becomes clear that the collaboration system is central to the project or programme's decision-making processes.

Another buy-in factor will be the interpersonal skills of the vendor's staff. Not all IT staff are good at thinking about the needs of the user; they may prefer sitting in front of a computer screen and interacting through that. Good interpersonal 'chemistry' between the vendor's staff and other team members will certainly help (ideally, the vendor might become accepted as another project team member contributing to the successful completion of the project in question). Their ability to identify opinion formers and/or senior team members to act as 'champions' can also be helpful.

8.3.4 Agreeing exchange standards

Trying to maintain the sense of involvement hopefully already created by including end-users in the selection of a system, most vendors will organise workshops where the focus shifts from demonstrating the system's features to discussing how the system will be used on the particular project or programme that is about to commence. Also as mentioned in Chapter 5 and in Chapter 7, a key stage is the collective compilation of a project- or programme-specific protocol document detailing the team's processes, its deliverables and the technologies and information exchange standards it will use.

In the early days of construction collaboration technology, vendors tended to work with project teams to develop protocols that would be specific to the project and to their particular application. However, particularly where team members work on several different schemes for different clients and use different collaboration systems, the process of generating a protocol can often be repetitive and inefficient. Recognising that the growth of web-based construction collaboration technologies was exposing an increasing number of UK clients and their project teams to the challenge of agreeing appropriate structures and standards for electronic information exchange, two government-backed bodies developed guidance. In 2003, Construct IT published 'How to Manage e-Project Information', a simple guide that provided valuable checklists to help teams identify what content they needed to capture, and the sources of that content. It also discussed how teams might establish information processes, but devoted only a page to this. Fortunately, the other government-backed project was about to fill the gap.

In 2002, the Building Centre Trust had embarked on a DTI-funded project to develop an industry-standard PIX Protocol. The first PIX Protocol guide and toolkit was launched in March 2004.[7]

The PIX Protocol's forms and checklists cover most of the information collated during generation of project- and application-specific project protocols. Whether developing a protocol using the PIX toolkit, using a vendor-specific protocol or using one developed by the client or another team member, end-users need to look closely at several issues governing the smooth running of a project. The range of topics extends from what IT infrastructure, software applications, document and drawing systems, email distribution policies and internet usage policies end-users currently use *individually*, to what formats will be used for each type of information, what drawing status controls will be employed, what document naming/numbering conventions will be used, etc, by them all *as a team*. The following are just a few extracts from the Protocol's 'Project leader checklist':

> Project leader's agenda at project initiation:
> 5.   '...as soon as the core team members are appointed, initiate a meeting to agree, test, publish and implement the PIX Protocol for the project.'
> 6.   'Ensure that the client implements the agreed PIX Protocol in all subsequent appointment contracts to the project.'
> PIX Protocol meeting agenda:
> 1.   Establish Information Exchange Principles...
> 1.4 'If editable documents and data models are to be exchanged, agree the legal status for each format that is to be exchanged, determine the responsibilities of each party to the exchange and the verification procedure for the data being exchanged electronically.'...
> 2.   Establish Design Management Principles...
> 2.6 'Agree the coordination process and responsibilities for review, commenting and approval of drawings and documents within the design team and for those produced by suppliers and subcontractors.'
> 3.   Establish Document Management Principles...
> 3.4 'Agree security status for each document and agree who should have access to the document'
> 5.   Project Communications...
> 5.3 'For electronic communications, consider adopting an Electronic Data Interchange Agreement.'
> 5.4 'Agree the status of email. In particular differentiate between "contractual" and "conversational" email.'
> 5.5 'If the project is to have an extranet or other type of network solution, agree roles, responsibilities and procedures and particularly folder structures for storing project information.'
> 5.6 'Agree minimum standards for firewall, virus protection software, back up and antivirus software updating procedures.'
>
> (Building Centre Trust 2004)

Such attention to detail is important. For instance, agreeing distribution lists can help streamline a project and avoid confusion; avoiding a 'scatter-gun' approach and targeting information on those who need to know can help prevent interference by other team members; the latter may also appreciate not being bombarded by information about matters outside their project remit. Perhaps more importantly, accurate targeting of information can also buttress contractual roles and responsibilities, by ensuring, for example, that all the parties needed to finalise an element of the design are included on a distribution list or invited to contribute to a particular collaborative process.

8.3.5 *Training*

Ease of use (see Chapter 5) can also be a critical factor in encouraging take-up of the systems. As most of the systems are web-based, their use of common website navigation features such as hypertext links and clickable icons will quickly overcome some resistance, and proficient IT users may even start using the systems with minimal training. But it is naïve to think that, however intuitive to use the vendors claim their systems to be, that new users will be able to start using a system effectively without some initial training or support. Appropriate training is vital.

In many instances, the introduction of construction collaboration systems will often alter existing working practices. For example, a design task may no longer conclude with the printing, copying and distribution of multiple copies of a paper-based drawing; instead, the CAD file may be saved into a different electronic file format(s) and simply be published to the central repository in one operation. Similarly, issuing instructions, RFIs, change orders and other project notices may now be undertaken electronically without the need for senders and recipients to record their progress in local systems.

Training in how to use the technology will vary according to the different roles and responsibilities of individual end-users. For example, within a project team, there may be designer users whose work will involve frequent publication of and commenting upon drawing files, project manager users may be more concerned with monitoring and managing particular detailed processes (e.g. RFIs), while there may be more casual users who simply need to be able to access the system occasionally and view only a few items of information. The extent and content of training will therefore need to be moulded to the needs of different groups of end-users, and will need to be staggered to accommodate the different times at which staff are mobilised to work on the project in question.

The introduction of construction collaboration technology to a project team can also mean new project roles and responsibilities for some end-users. For example, within individual companies, it may be necessary for someone to act as an administrator and provide first-line training and support for colleagues who become new users; within wider project or programme teams, someone may need rights to change system settings, set up new projects, grant access to new team member companies, etc. Some providers work on a 'train the trainer' basis – an approach supported by research evidence:

> training is vital to ensure the success of …implementation. At a minimum, everyone who will be using [the system] in a project needs to be trained on how the system works and how it relates to the team's business process early in its implementation.

> Although a consultant or service provider may be used to help during the implementation process, it is important to ensure that knowledge is really transferred from the consultant or service provider.
>
> (Nitithamyong and Skibniewski 2004, p. 21)

Again, the extent and content of training will differ according to whether individuals take on such roles.[8] Teams should exercise care in appointing a project coordinator or administrator; these individuals need good levels of understanding of construction processes and of the technology to be used, and should also have good people skills so that they communicate well with and command the respect of fellow team members, and help build buy-in to the system's use.

IT-literacy will obviously also be a factor. Training requirements will vary according to the trainee's previous familiarity with computers, with using an internet browser, or with other collaboration systems. The Construction Industry Council's professional services skills survey (2004) found there were gaps in general IT and professional IT skills among 11 and 14 per cent of existing staff respectively, but new recruits were often more able: general IT skills were lacking in just 2 per cent of cases, while professional IT skills were lacking in 9 per cent of applicants. Echoing the already-mentioned distinction between different sizes of organisation, the CIC also found 'professional IT skills are a larger problem in smaller firms [and] general IT skills are a larger problem among larger firms' (p. 26). The picture may change, however, and some educational institutions are beginning to teach students about online collaboration. For example, postgraduate project management students at UMIST in Manchester and at the University of Greenwich have been addressed by guest lecturers from UK system vendors, while the University of Newcastle included hands-on use of the BIW system as part of a five-month module in its MSc in Digital Architecture.

Previous experience with a collaboration application can also improve performance (Nitithamyong and Skibniewski 2004 – they also found that CAD experience significantly contributed to performance) and reduce the training requirement, but may not remove the need altogether. While the systems provided by the main UK vendors may have many similarities, they also have many differences: near identical features can have different names and function differently, and, of course, project protocols may differ. Even after previous experience with the same application, some users may still need 'refresher' training, perhaps to bring them up-to-date with new functionality added since they last used that vendor's system. Perhaps the most seamless introduction occurs when a particular system has been adopted as a standard platform for project delivery; project team members – perhaps working within a long-term framework contract to familiar project protocols – simply just start working on the next project.

8.3.6 Cost

It is an indication of the cost-consciousness of UK construction industry staff that the first question many people ask when they talk to collaboration technology vendors at conferences or trade exhibitions is: 'How much will it cost?' to which the stock answer was normally something like: 'It all depends...'. Such imprecision is not surprising. The costs can fluctuate depending on how the software is licensed, on who is paying, on the size, complexity and duration of the project or programme (which have a direct bearing on numbers of users, storage, etc.) and what stage it has reached, on training and other implementation requirements, on depth of required functionality, etc.

As detailed in Chapter 3, the main distinction is between customers paying a large, up-front perpetual or term licence fee – usually so that they can host the application themselves and based on a certain number of users or 'seats' – or paying smaller fixed monthly subscriptions to rent the software on a remotely hosted ASP basis to use as much as they want (i.e. no limits on the number of users or documents, or on storage capacity) for as long as they need it. Most of the leading construction collaboration providers active in the UK market have opted for the latter model. The precise cost is normally negotiable, but will tend to reflect the size, value and/or scope of the capital project or programme concerned. These metrics help indicate the number of users, the volume of information to be exchanged and stored, the extent of functionality required, configuration, implementation, training and support needs, etc. Reflecting this, the subscription costs will typically also vary according to the different stages of a project; an initial rate may be charged for the pre-construction phase, but the rate is increased for the duration of the construction phase, after which the rate may return to a much lower rate if it used to cover FM, operation and maintenance, etc.

There will be many projects where use of sophisticated construction collaboration technologies may not be justified. For example, it would plainly be inappropriate to employ them on minor building works schemes of short duration undertaken by small teams with little or no requirement to use IT solutions such as CAD. Teams should weigh up the costs of procuring and setting up the system and training the users against the likely benefits (see Chapter 9); generally speaking, the larger, longer or more complex the project or the larger or more far-flung the project team, the more likely it should be that a collaboration solution could be deployed. Some of the UK vendors suggest minimum project values (rules of thumb vary from £0.5 million to around £2 million) below which use of their systems may not be feasible, but there can be exceptions. For example, where a client has adopted a particular system as its corporate standard and its supply chains members have become proficient users, it will cost little to set up a new project and no additional training costs will be incurred; in such circumstances, some clients have used their system to support projects worth less than £50,000.

Who pays? Early UK adoption was driven by some prolific industry clients with substantial programmes of work and a strong partnering ethos (see Chapter 2); they selected and paid for the application and mandated their project teams and supply chains to use it. More typical 'occasional' clients tended to learn about online collaboration from members of their professional teams, and once persuaded of the technology's value, might either pay for the application themselves or agree that one of the project team could include the collaboration system as part of its services and include its costs within its fees. Once it becomes clear that other individual end-user companies would not have to pay for the application themselves, a major obstacle to its wider adoption and use is immediately removed, though it also helps to inform prospective team members as early as possible:

> A key lesson from the project is that if such a system is to be included all parties must know at tender stage. 'I think it's imperative that all firms that bid for a project know that they are expected to go online. That way they can cost whatever software and hardware requirements they might have into the tender. It's no good a specialist contractor finding out after the event that the project is not a conventional paper-led one.'
>
> (Subcontractor quoted by Faithfull 2003)

Some clients, consultants and contractors have moved beyond the 'pay-as-you-go' model and – echoing the partnering model already extensively used within the UK AEC industry – have signed longer-term contracts with providers, whereby they commit to manage all their projects using that provider's system for, say, a five-year period. From the customer's perspective, such strategic relationships can result in major savings as they can negotiate discounts for large numbers of projects,[9] and integrate the collaboration application more closely into their business processes; the provider, on the other hand, gets greater certainty of future workload and income.

At the opposite extreme (and reflecting again the predominant 'short-termist' industry focus on cost and immediate return), many customers want to pay as little as possible, and many of the early buying decisions were taken almost solely on the grounds of cost, not best value, with some providers looking to undercut their competitors in order to win market share (sometimes at price levels that were unsustainable). As in many walks of life, cheapest is not always best; some customers subsequently found that the functionality and performance of their chosen solution and/or the quality and reliability of their vendor's services tended to reflect the lower price paid. As already discussed, even detailed pre-qualification procedures will be unable to investigate the vital human interaction necessary to the successful delivery of construction collaboration services. As a result, with lowest price the key determinant, industry customers (and their supply chains) may find themselves dealing with businesses who are ill-equipped to collaborate or to enter into long-term relationships, or who perform badly on other key metrics such as service quality, health and safety, environment, etc.

Perhaps the most notorious example of cost-consciousness is the reverse auction (where pre-qualified suppliers compete to offer the lowest price to win the contract). During 2003, a major UK contractor employed a reverse auction process to procure collaboration services, and was criticised by some participants. Providers argued that the pre-qualification questionnaire was insufficiently detailed, and capable of wide interpretation; most vendors could tick all the boxes even though their justifications would have been quite different. The customer brief was complex and there was considerable scope for negotiation about what key elements would be delivered, when, how and by whom (remember: successful collaboration is only 20 per cent technology, but 80 per cent people and processes).

Making price the key selection determinant may:

- *Lead the vendor to compromise on delivering a quality service* The most competitive price may mean corners need to be cut.
- *Lead the customer to compromise* The best scorers on technical, managerial and business-related issues may be discounted if the main selection priority is price.
- *Increase project risks* IT is not a simple commodity. Procuring IT services mainly on price poses huge risks. Poor, unreliable or inadequate systems may be procured, requiring huge additional investment to rectify later problems, and increasing the risks of disputes and costly litigation. In short, the initially cheapest can become a long-term 'money pit'.
- *Reduce research and development* IT is a sector that invests constantly in research and development. Price-cutting to win work runs the risk of a knock-on effect, curtailing the ability and/or willingness of vendors to continue to develop their systems. This in turn will hamper the efficiency and effectiveness of the teams (and, in due course, the industry) using the technologies supplied.
- *Reduce choice* Heavy AEC reliance on price-driven processes such as reverse auctions may lead some IT vendors withdrawing from the sector.

What additional costs might be incurred? So far in this section, the focus has been on costs associated with software licensing, implementation and training, but – depending on the telecommunications, hardware and software already in place – the introduction of construction collaboration technologies may have an additional financial impact on some members of the supply chain. For example, there may be still be some businesses that have yet to start exchanging information via the internet: new internet connections and perhaps new computers may be needed. Removing the designers' burden of printing, copying, folding and posting paper-based drawings[10] may simply mean that printing and paper costs are passed down the supply chain to contractors, specialist sub-contractors and suppliers, who still need to print items off locally (see Section 9.3). Such requirements should be noted during the compilation of project protocol documents so that these particular end-users can make economic use of their existing resources, can upgrade them as necessary at modest cost (e.g. agreeing on use of A3 drawings so that suppliers need only invest in an A3 laser printer rather than a more expensive plotter) or work around the issue (e.g. sending some drawings to a bureau service for local production of prints where required).

8.3.7 Differentiation or competitive advantage[11]

If US experience is anything to go by, customers will increasingly begin to demand that their supply chain partners are web-literate. By 1998, several major American building owners and clients (e.g. Nations Bank and 3Com Corporation) were already demanding that their consultants and contractors become internet-literate just to remain qualified to bid for projects. Anticipating that the UK construction clients would follow suit, some consultants, contractors and suppliers began to work with the various extranet technology providers, even setting up operations that were focused specifically on offering a complete solution. For example, some professionals, particularly those (e.g. construction managers) with responsibility for delivering schemes on behalf of their clients, looked to combine their traditional strengths with expertise in advising upon, implementing and supporting collaboration technology.[12]

Such businesses add to their value propositions, being able to offer IT audits, consultancy, training and other web-related skills that complement their core services. Moreover, by advocating collaborative technology, they achieve other marketing advantages:

*   *Market positioning as innovators* Championing the Latham and Egan principles of partnering, lean construction and integration;
*   *Better processes* Use of collaboration technology can help improve the solution partner's own internal processes, and improve their management of the project and the client's supply chains;
*   *More competitive pricing* By being able to accurately gauge the cost savings that can arise from using the technology, the consultant/contractor can also price its services much more competitively.

Duyshart (1997) summarised the potential advantages to be gained from adopting new IT:

> significant gains can be made by those professions which adopt the use of digital document technologies and leverage advantages of their use to achieve aims such as increased accessibility to information, improved operational performance, improved quality of product, development of more efficient methods of practice, reduced production and distribution costs, savings in travel and disbursements, and increased value from information repositories and knowledge bases. Practices which adopt these technologies early, and use them as a means of improving the quality and efficiency of their services, stand to gain a significant positional advantage in the industry.
>
> (p. 200)

Not adopting the new technology may also pose risks, of course. The opportunity was memorably summed up at a Construction Information Technology Alliance seminar in Dublin in May 2004 as offering the 'potential to streamline or sideline your business'.

Once an organisation can genuinely demonstrate that it is has the right combination of people, processes and technology, it is then in a position to become a more permanent and integral part of a client's delivery process for whole series of projects. This may be a precursor to more fundamental changes both within and across supply chains.

Through being web-literate, there are also opportunities, perhaps, for architects and other designers to either re-assert or to protect their positions as controllers of projects. For many years, they retained a central role in the development of new capital assets by acting as a producer, manager and disseminator of paper-based project drawings and documentation. For architects, says Cohen (2001, pp. 253–254), internet-based technology may be the means by which the profession reasserts a central place in the direction of projects by assuming an important new role – that of project information manager.

Greater use of collaboration technology may also lead supply chain SMEs to rethink how they respond to customer opportunities. They might, for example, form temporary consortia or joint ventures – 'virtual companies' (Egan 1998) or 'networked organisations' (Cohen 2001, p. 254) – or even merge their operations permanently, pooling their resources so that they can collectively respond more efficiently and effectively to their customers' demands. On their own, many small contractors, sub-contractors and consultants may be unable to provide the required level of service. But by combining with other enterprises with complementary skills and/or resources, and using collaboration technology to share data, they could become valuable members of the client's team. To some extent, this has already begun to happen. As mentioned in Chapter 2, PFI/PPP teams have been using systems to manage the vast array of documents and drawings required to support a successful bid. With a project's whole-life costs increasingly important, both PFI/PPP consortia and traditional clients have also started to look to their extranet systems to manage information beyond the design and construction phases, throughout the life-cycle of their facilities. And as design, construction and operation factors cease to be differentiators, the creative nature of collaboration may help consortia develop new ideas that give them a marketing edge. For example, the Bucknall Austin-led prime contracting team that redeveloped the MOD's facility at Andover North (ITCBP 2003b) subsequently created a new organisation, Novus Solutions Ltd, to target other customer opportunities demanding similar skills and experience.

8.4 People, processes and technologies

Throughout this book, it has repeatedly been emphasised that successful collaboration requires a combination of people, processes and technologies. Moreover, as this chapter should have made clear, it is often the first item – people – that is the most important and the most difficult to get right.

Prospective users of construction collaboration technologies need to be alert to the resistance that many people within the UK construction sector still have to the basic idea of collaboration. From individual 'creative isolation' attitudes, through interdepartmental turf battles and inter-company distrust, to wider industry conservatism, there are many factors that can militate against project teams being able to work together effectively. Even assuming one is able to overcome resistance to the notion of collaboration, one might still need to the address a wide range of practical issues that affect the ease with which teams adopt and use collaboration technology successfully.

> The expectations of a collaboration solution promote an environment where individuals actively share ideas and information freely in a fluid and dynamic fashion. However, the reality is that we are all driven by the basic premise of 'what's in it for me?' and people are generally not willing to share information. There is simply too much conflict, political jostling, and self-interest.... Organisations need to be cognisant of this when implementing collaborative technologies. Thus, where possible, collaboration needs to be built into the organisation, its business processes, and applications, in order to ensure active participation. The reality of overcoming the paradox is grounded in the alignment of people, the business requirements, and the technology.
>
> (Butler Group 2003)

8.5   Chapter summary

This chapter has drawn together the threads of the five preceding chapters and explained how each of the topics (the technology providers, hosting issues, the technical capabilities of each software application, internet connectivity and legal concerns) might be raised quite legitimately as reasons why construction collaboration technology might not be introduced and used on a particular project or programme. It has then moved the discussion forward and focused on the critical human element, aiming to help the reader identify some of the issues that, if successfully addressed, can facilitate the introduction and use of the technology. Chapter 9 looks at perhaps the single most important positive influence on use of collaboration systems: the benefits they deliver.

Benefits of using construction collaboration technologies

This chapter:

- summarises the main tangible and intangible benefits claimed for construction collaboration technologies;
- looks at the emerging evidence of those benefits, yielded by vendors' case studies and by independent research;
- discusses how project teams might measure the return on investment (ROI) of using the technologies.

Many AEC industry clients and their supply chain partners have been sceptical about the claimed benefits of construction collaboration technologies and have therefore been slow to adopt them.

> The primary motivator for actors in the AEC industry to adapt new technology innovations will always be the opportunity for direct gains and benefits in their own operations. In order for the actors to realize these benefits there must be a framework in place to measure the relevant cost and benefits associated with the investment. Lack of any quantitative study by an objective party on Return on Investment (ROI) has made it difficult to justify cost/benefit.
>
> (Becerik 2004a, p. 10)[1]

However, this picture has slowly begun to change. All of the main UK providers can now provide details of clients and projects that have used their services, allowing prospective customers and project team partners to seek references and to read case studies; browsing through the websites of the main providers, numerous advantages and benefits are claimed for both team members and client organisations. This chapter will summarise the main areas, looking at both tangible and intangible advantages, and also outline some of the disadvantages. It will then review some of the independent research and the vendor case study evidence, before outlining the basis by which potential customers might make an informed decision based on a calculation of the ROI involved – assuming, of course, that team members are prepared to record the relevant measurements.

9.1 Tangible benefits

The most frequently claimed savings tend to relate to direct, tangible or quantifiable benefits such as cost reductions and time savings. Few vendors will claim that employing their technologies will immediately make the project cheaper or that it will mean earlier project delivery, but there are frequent claims that they will reduce many of the expenses typically incurred and speed up many processes that were slow primarily because they were previously mainly paper-based. For example:

- reductions in expenses, materials and man-hours relating to printing, reproduction, distribution (including post and couriers – internal and external), storage/archiving, management and retrieval of drawings, documents, photographs, forms, etc.;
- reductions in travel, meeting and fax/telephone costs;
- less time spent searching for or chasing already-existing information, or working on out-of-date information;
- faster drawing revision cycles and other approval processes.

There is still considerable scope for industry organisations to achieve these benefits and more. Autodesk's 2003 survey of the UK AEC industry found that, while 86 per cent of its respondents said they used IT to communicate regularly or most of the time, for over 60 per cent this means the regular use of CD-ROM and post to distribute data to third parties. While 20 per cent made use of the internet or an extranet facility to support data distribution, the dominant modes remained email and/or CD-ROM by post, with nearly 30 per cent still using couriers regularly to transmit data. Worryingly, a startling 16.2 per cent of time was said to be wasted through the use of outdated design data (i.e. more than an hour every working day).

Reading the vendors' own case studies yields numerous anecdotes about the tangible benefits achieved on paper-related costs, for example:

- over £58,000 savings on printing, copying and postage during a £5 million, 30-week retail project (BIW);
- a £50,000 saving on printing, information distribution and general administration on a £60 million development (BuildOnline);
- a £12,316 saving on a £1 million project (Cadweb);
- a potential £27,250 net saving on a £10 million, two-year project (Causeway).

Similarly, time savings could also be recalled. Seamus Mockler of Kajima told a 2004 London conference:

> Traditionally, in the UK it can take up to 10 days for drawings to be reviewed and then re-issued. By using BIW Information Channel we were able to reduce the drawing review process to an average of two days and to half a day at one point.
>
> (quoted in ITCF Newsletter July 2004;
> see also Section 9.6.1)

Another construction manager worked on a building for financial corporation CapitalOne Bank:

> A contractor published a drawing of a mechanical item for comment at 8.30 am one day recently. Sixteen people commented on it during the morning. The contractor uploaded a revised drawing at 4 pm that afternoon – again for comment. The entire matter was resolved in less than two days. If this had been done by printing and posting, it would have taken a week or more.
>
> (BIW case study)

9.2 Intangible benefits

It is, of course, more difficult to document savings relating to soft, indirect, intangible or more qualitative benefits. The claimed benefits can again be roughly divided into the three key risk areas of time, cost and quality (though there will inevitably be overlaps where, say, a delay adds to the costs incurred).[2] For example:

Cost

- fewer claims for lost, incorrect or out-of-date information (audit trail encourages accountability and adherence to programme);
- less re-working;
- less reliance on paper (expensive to generate, distribute, store and retrieve);
- fewer disputes and litigation;
- better control and forecasting of project-related cash flow;
- lower IT costs (overheads for email, firewalls, etc., plus IT staff time, can be discounted).[3]

Time

- faster mobilisation of initial team members; subsequent new joiners can get 'up to speed' more quickly;
- greater flexibility (anyone with computer and internet access can use the system; no need for specialist hardware or software);
- faster access to drawings, documents, etc., following issue (e.g. transaction potentially completed in seconds online instead of hours or the following day if couriered or posted);
- earlier/more timely involvement in key decisions (concept, planning, surveying, design, specification, fabrication, construction, installation, maintenance, repair, replacement, etc.);
- less time wasted searching for information;
- faster communications (supplier-specific decisions communicated more quickly and completely; streamlined processes, for example, less time spent processing RFIs,[4] change orders, etc., fewer RFIs, change orders, etc.);
- more information is shared immediately electronically instead of being converted to paper and being scanned or re-keyed;
- less employee time spent on administrative work, including form-filling;

- greater employee productivity (e.g. project managers can manage more projects);
- more convenient information sharing – employee mobility no longer an obstacle;
- fewer drawing revisions;
- reduced design work (from using the system to maintain and manage corporate design standards, for example);
- fewer unnecessary project delays or earlier identification of mistakes or inconsistencies; faster problem-solving and elimination of distribution, postal delays or other bottlenecks;
- faster compilation of project hand-over information (e.g. Health and Safety File);
- earlier completion dates due to time savings in transferring key information (resulting in lower on-site costs and earlier revenues to owner/developers through use, rental, lease, etc. of the facility);
- faster evaluation and resolution of claims.

Quality

- fewer mistakes;
- avoidance of doubt through 'a single version of the truth': all data is stored on one system for all authorised project members to see; core information (e.g. project team contact details) available to all;
- fewer data compatibility issues (e.g. no need for non-CAD users to have CAD software);
- less 'information overload', most notably through reduced reliance on email;[5]
- better collaboration: more open, cross-discipline discussion and coordination of design issues (users can review, discuss, mark-up, and ask and answer questions about each others contributions), leading to...;
- improved understanding of project and processes (better design, less duplication, fewer errors and less re-work), better problem-solving (e.g. a joinery contractor could look at the electrical design so that his work took account of cabling, etc.) and better decision-making;
- greater transparency (e.g. reporting tools can be used to summarise outstanding actions to be resolved at project team meetings);
- better reporting and tracking or auditability (audit trails detail who did what and when);
- increased scope for creativity and innovation (online collaboration may stimulate new ideas);
- better implementation of and adherence to corporate standards;
- improved monitoring of individual professionals' and companies' performance;
- better customer/supplier relationships;
- greater re-use of information within a project (less 're-inventing the wheel', more WORM – write once, read many);
- as-built data and associated product information becomes part of knowledge base for future projects (or repeat aspects of the same project) and part of operation and maintenance and health and safety systems;
- more re-use of standard information across a series of projects – information is not dispersed along with the team members after project hand-over;
- more resilient, reliable, robust and secure data management infrastructure;
- improved levels of IT literacy among project team members.

In addition to the above benefits, one might also highlight the social and environmental benefits to be gained from reducing the volume of paper distributed, cutting the amount of travelling undertaken, allowing more flexible working patterns, etc. On its website, for example, Cadweb highlighted reductions in energy consumption and pollution resulting from reduced deliveries by reprographics firms to its projects, and developed a 'football pitch report' to describe the potential paper savings arising from using its collaboration system to issue drawings (equating the amount of paper saved to the equivalent area of several standard football pitches). Similarly, in November 2004 BIW calculated that users of its system had distributed electronically enough drawings to build a pile of paper over a kilometre high. In an industry often maligned as low-tech, other indirect benefits might include an improved reputation with clients at both company and industry levels.

9.3   The disadvantages of using construction collaboration technologies

The vendors will naturally be very positive about the advantages to be gained from using their systems, but some supply chain members suggest that there is downside to the use of the technologies. For example, if we revisit some of the cost, time and quality issues:

Cost

- Additional costs to purchase or upgrade telecommunications links and/or IT hardware to access and use collaboration systems efficiently.
- Additional supply chain costs to purchase, lease or upgrade drawing plotters or printers as drawings are no longer received in paper form[6] (many users may still prefer to print out a drawing for commenting purposes (Croser 2003), arguing – with some justification – that it is not easy to navigate around and mark-up, say, an A0 drawing on a 17-inch computer monitor. Such issues can partly be addressed by, for example, agreeing project protocols about drawing sizes, while vendor feedback does suggest that willingness to review drawings on-screen increases as end-users become more familiar, confident and proficient at using such tools).
- Following on from the last point, additional costs of printing consumables simply passed down the supply chain.
- Electronic communication makes it easier to revise a drawing and may result in issue of more drawing revisions.

Time

- Slower mobilisation of initial team members as they must first agree project protocols and, if necessary, get training in use of the chosen collaboration system.
- Additional training needs for subsequent new joiners.
- End-users may need training in multiple systems where, say, their business is working on several projects each employing a different system.
- Some system processes can seem very time-consuming (Croser quotes an architect: 'Sometimes I just set a search going and make a cup of tea'), though this

tends to be an IT infrastructure issue and, in any event, the end results – for example: document histories, audit trails, etc. – are far more efficiently compiled than would be possible by manual methods.

- Multiplication of project communications – participants may adopt a 'belt and braces' approach to communication, sending information to all project participants instead of selecting just the relevant participants.

Quality

- Too much collaboration – allowing wider and easier access to information can allow more people (some of whom may be unqualified) to offer sometimes unhelpful comments, ideas, etc.
- Too transparent: 'Big Brother' can monitor individual professionals and companies' performance, and – while this may be a major benefit to project managers and clients – it may be too transparent for some individuals or businesses, particularly those who have yet to embrace more open or partnering-oriented approaches to project delivery.
- Only truly effective if all project team members use the system.
- Loss of interpersonal contact or interaction (through fewer meetings or telephone conversations).

In most cases these issues are relatively short-term and not insurmountable, but they re-emphasise the need for teams to take a planned, detailed and pragmatic approach to the human aspects of the introduction and use of construction collaboration technologies (such matters were considered in Chapter 8). While a supply chain company may claim that the introduction of construction collaboration technology has necessitated investment in new IT hardware or software or upgraded telecommunications, it is often found that it such steps would probably have been taken eventually – the need to collaborate simply brought it forward – and the equipment will probably be used to benefit other projects.

9.4   The technologies' short track record

Track record is, rightly, highly valued in the conservative and risk-averse AEC industry. After all, in many instances, the client will be entrusting the delivery of a capital asset that may be worth millions of pounds to a team about whom they know relatively little. Understandably, the prudent client is going to seek references from previous clients of potential team members, and will look closely at the previous achievements of that team both at individual and company levels. It follows, then, that the same care should be applied to the selection of a technology and its vendor to support communications between those project team members. But by the mid-2000s, construction collaboration technologies were only slowly becoming a normal part of the project delivery process, and so had established only a limited track record. There were several reasons for this:

- *The relative newness of the technology* As already outlined, the technology tended to be adopted first by 'innovator' clients, contractors and consultants; there was still some way to go before it overcame the suspicions of less progressive/more pragmatic individuals and organisations. One consequence of this was that the total user community remained relatively small: any research surveys could only be based on small and unrepresentative samples.

- *The immaturity of the market* Again as already mentioned, the technologies suffered from being associated with the boom and bust of the dot.com era. Buyers were wary of committing their projects to vendors with short track records and/or uncertain futures, to systems or architectures that might not prove to be 'industry standard', and to internet-based solutions over which – as they saw it – they had less control.

- *The long duration of construction projects* Most construction schemes take many months, even years, to move from initial planning stages through to completion, and with track record often a key factor in buying decisions, building a credible previous history can take time.

- *Adoption requires multilateral commitment* Unlike the unilateral adoption of a technology employed within a particular business (whether a generic application such as word-processing or a discipline-specific tool such as CAD or cost control), the decision to adopt and use a collaboration system tends to require pan-disciplinary commitment and involvement and, in many instances, the active leadership and encouragement of the client. Such consensus may not always be forthcoming.

As a result, much of the evidence is anecdotal, few providers (or their customers or project teams) have consciously set out to measure the benefits – the DTI benchmarking study (2004, p. 95) found construction businesses were the least likely of all sectors to measure the costs and benefits of technology – and there is still relatively little independent verification of them. Influential American analyst Joel Orr (2002, pp. 11–12) believed there were several reasons behind the lack of research confirming overall productivity improvements:

- To speak of 'productivity increases' requires careful measurement of productivity prior to implementation of the new technology. This is difficult to do in the project-oriented world of construction, and companies do not seem motivated to do it.

- Some issues are self-evident, and construction professionals do not want to invest time in proving them – for example, the fact that electronic transmission of documents...is much faster and cheaper, and more auditable, than using courier services....

- For many of the parties to a construction project, productivity is not a clearly defined concept. To put it bluntly, if one is being paid by the hour, reducing the number of hours required to get a job done is not an attractive proposition.... Only the owner is clearly motivated to do more with less....[7]

- Most...vendors underestimate the extent of computer-illiteracy in the construction community, and thus underestimate the amount of training required for successful project implementation.

- Construction projects are not highly disciplined affairs. Unless the use of a new tool can be tied to payment, sub-contractors will tend to do things 'the old familiar way', despite any benefits they might gain from the new tool.

Similarly, some industry professionals often insist that every construction project is unique, and that the benefits gained from one project will not necessarily be experienced on the next. Even though the differences may be quite minor, many industry professionals still tend to see each scheme as a one-off. They might argue that a project brings together different individuals from different companies, working on a different site in a different location with different access and service routes, developing different designs to different time and cost constraints, using different IT hardware and software solutions, and constructing the asset under different terms of contract for delivery to different types of customers or end-users with different approaches to maintenance and operation.

In the same vein, different benefits, or different levels of benefits, might be achieved through use of different collaboration systems. Some benefits might apply across all the main systems, but others might only be delivered by particular individual applications.

Some benefits might only be achieved because the project team also had an impact. It can be difficult to determine whether benefits have been achieved either solely or partly due to the technology or whether they were due to other unrelated factor(s). For instance, how does one quantify the benefits of using collaboration technology on a project with a strong partnering ethos? Team members may experience, say, a dramatic reduction in the volume of paperwork produced, distributed and stored. While this may reflect the use of the collaboration system, the reduction may also have arisen because the team culture had removed the need for the many contract letters often found on traditional projects.[8] Arguably, the only way to prove which influence was greatest would be to run two identical projects using the same team members – one using the technology and the other not – but the absence of a collaboration system may fatally undermine the open partnering ethos. Often the two may be interrelated. For example, Howard (2002) describes an interview with staff from airports operator BAA plc in which overall 3 per cent cost savings were achieved due to use of collaboration software and partnering framework deals. Some cost savings were attributed more to the common project team (cutting design team costs, shorter programmes), while others tended to be attributed more to the use of the technology (e.g. fewer drawing revisions, faster drawing approvals, reduced design work, greater transparency), while some were solely due to the technology (use of product library for standardisation, electronic health and safety file). Overall, 70 per cent of the project savings were attributed to the technology.

Perhaps more importantly – and getting to the heart of the construction industry's widespread blame culture – when track record remains a key differentiator, industry businesses will usually be reluctant to admit that they can sometimes be inefficient or make mistakes, that activities were duplicated, that tasks had to be redone, etc. Drawing on past experience, many project budgets and programmes are prepared with healthy contingencies built into them to allow for unplanned events, some of which often arise simply because of communication breakdowns between project team members. Detailed research into the use of collaboration technology on a project might expose such inefficiencies. Producing a case study which asserted that project X had been completed more quickly, within budget, and with fewer mistakes might suggest to the reader that the firms involved had previously been somewhat lax.

9.5   Independent research

As mentioned, apart from vendors' own case studies or surveys, independent research in the United Kingdom into the performance of construction collaboration technologies has been limited (more research is emerging in the United States and in Scandinavia). Moreover, the small number of UK studies undertaken to date have tended to be reliant on small, unrepresentative and therefore statistically unreliable samples and have focused on eliciting users' subjective feedback rather than attempting any measurement of the benefits themselves.

For example, Breetzke and Hawkins (2003b, see also Martin 2003) based their report on information from just 12 organisations (9 consultants – mainly surveyors – and 2 contractors) who, between them, had experience of using 16 different systems, and the findings were, as the authors admitted, at best anecdotal. The list of benefits suggested by respondents included many of those listed in Section 9.2, but there were no clear conclusions about time and cost savings – 'two respondents recorded estimates of up to 10 per cent saving on project duration'; 'a slight majority felt that no [cost] savings were made' – while 'the majority of respondents did not consider extranets saved [staff] resources'.

Murphy (2001) fared only slightly better with an international survey, getting 47 respondents (22 architects/engineers, 13 contractors, 4 clients and 8 'others'). Between them, 28 members of this sample had experience of using 24 systems, including both US- and UK-based applications, ranging from basic information repository systems to 'active involvement' extranets with project-specific interfaces and functionality. Only half of his respondents used the latter, his results did not differentiate between American and British experiences, and his survey did not make a detailed assessment of the perceived benefits. His survey suggested: over 80 per cent felt the communication solutions offered 'improved administration'; extranet impact on project scheduling was not significant; over 60 per cent felt the solutions improved the resolution of problems or variations; and 96 per cent of respondents 'would certainly use a construction project extranet to host future projects'.

Murphy also undertook a small series of interviews and, again, the conclusions, while positive, are only anecdotal. The perceived benefits also included many of those listed in Section 9.2, for example reduced and controlled direct costs such as printing, courier and other office overheads; improved management and administration; and improved accountability. The auditable paper trail was considered a particularly powerful tool through both elimination of litigation discovery costs and the deterrent effect (significantly reduces the likelihood of a claim being implemented).

The results of a survey undertaken in late 2002 by Chien (2003) provided a ranking of the most noticeable benefits gained from using 'project extranets'. In order, the most important benefits were felt to be:

1   Providing up-to-minute project information
2   Improving distribution of information
3   Providing a collaborative environment
4   Improving team communication
5   Storing knowledge permanently
6   Increasing design process efficiency

7   Saving cost and time
8   Increasing productivity
9   Reducing the risk of errors and rework
10 Reducing the risk of dispute and litigation
11 Project on time or early completion.

At first sight, the findings regarding cost are perhaps disappointing, but Chien said many of his respondents 'emphasised that cost benefits are not fully realised yet and that it is difficult to predict cost savings at this early stage'. When asked to estimate what proportion of the total project cost saving can be attributed to using an extranet: 50 per cent put the figure at under 1 per cent, 23 per cent put the figure in the range of 1–2 per cent, 10 per cent put the figure between 2 per cent and 3 per cent, 7 per cent said 3–5 per cent, another 7 per cent put the figure between 5 and 10 per cent, and 3 per cent estimated the saving at 10 per cent or more.

In the United States, Nitithamyong and Skibniewski (2004) assessed the relationship between the use of a collaboration system and a construction project's success (measured by performance on cost, schedule, conformance to technical specification, health and safety, and owner satisfaction). Drawing on responses from users involved on 82 projects, using 14 different systems, they found strong correlations with only the cost and time parameters (p. 37).

Also in the United States, Becerik's survey (2004a) sent to 400 participants in three projects found respondents readily identified such benefits as improved communication, reduced printing, mailing and faxing costs, and reduced RFI and submittal turnaround time. Users were less confident that they would enjoy the benefits of reduced claims and litigation costs, increased competitive advantage or reduced change order costs, but the overall effect on project performance was perceived by a large majority to be either positive (46.5 per cent) or extremely positive (15.19 per cent); with 33.12 per cent rating the effect neutral, only around 5 per cent felt the impact had been negative. On a similar note, Bjork (2002), found that users of project webs in Finland would be very reluctant to go back to older ways of managing documentation: 'This can be seen as indirect evidence that the benefits are greater than the cost or possible inconvenience' (p. 6).

9.5.1   The ProCE project

Perhaps the most detailed and fascinating research, to date, into the benefits of using construction collaboration technologies was that undertaken by Sulankivi et al. (2002). This American–Finnish research project (also known as ProCE – Project Management and Organisation in the Concurrent Engineering Environment) developed a useful measurement framework to evaluate the benefits achieved during the processes relating to a single multi-partner project. The framework covered three categories of benefit: monetary benefits, other quantifiable benefits and qualitative benefits, and sought to measure these as they arose during different project-related activities:

•   input – communication;
•   control – document management, time management, cost management;

- execution – marketing, design (an extract from the framework relating to design is shown in Table 9.1), construction, operation and maintenance;
- resources – procurement;
- result – product/building.

As the extract in Table 9.1 illustrates, the ProCE team were able to identify a range of quantifiable and qualitative benefits relating to each area of project activity. In total, the framework included 15 potential 'monetary benefits', 34 potential 'other quantifiable benefits' and 47 potential 'qualitative benefits'. Looking at four sample projects (one each in Finland, Sweden, UK[9] and USA), the ProCE team compared traditional procedures with those employed using what they termed a 'CE environment'; combining subjective and objective measures, they administered interviews and questionnaires to project staff, and gathered data and documents from the collaboration solution. Benefit realisation was measured on a six-point rating scale from -1 to $+4$, and 27 of the potential benefits were positively identified by respondents (a larger sample, perhaps spread across more case studies, may have increased the total number identified). Space precludes a detailed consideration of all of the

Table 9.1 Example extract from ProCE benefits framework

Activity	Design	Monetary benefits	Other quantifiable benefits	Qualitative benefits
Execution	S = design	S11 Reduced labour and administrative costs (locating product information faster) S12 Cost savings due to fewer design errors (less rework)	S21 Time savings due to fewer design errors (less rework) S22 Improved quality of design (fewer design errors, improved constructability) *S23 Faster design change process* S24 Fewer administrative tasks *S25 Faster decision making* S26 Faster specification of building products and materials (eCommerce and other info sites) *S27 Time savings from reusing information more effectively (e.g. sharing electronic building layout)*	*S31 Better tools for collaboration improve collaboration possibilities* *S32 Easier to share common electronic information (e.g. building layout)* *S33 Improved coordination of geographically dispersed resources* *S34 Increased awareness of task status* *S35 Easier to publish updated/revised documents* *S36 Increased accessibility to up-to-date project information* *S37 More complete and consistent recorded and retrieved project information* *S38 Easier to monitor design progress (e.g. for design coordinator)*

Source: Adapted from Sulankivi *et al.* 2002, p. 43.

Note
The benefits identified in case studies are highlighted with italics text.

Table 9.2 Summary of overall ProCE case study results

Case no.	Qualitative benefits (overall score in scale −1 to +4)	Time and other quantifiable benefits	Monetary benefits	Monetary benefits/ cost of use
1 (Finland)	2.6 'The tool substantially helped in document and project management'	• Time savings: about 200 hours (29 work days) • About 1,700 days less information distribution delays	Altogether about $17,300	2.6
2 (Sweden)	2.2 'The tool was somewhat helpful in document and project management'	• Made about 4 months tighter schedule possible • Work time savings: about 50 hours (about 7 work days) • Time delays in the information distribution: about 530 days less • Fewer disputes, related to the information distribution	About $8,100	1.8
3 (UK)	2.3 'The tool was somewhat helpful in document and project management'	• Time savings: about 162 hours (about 20 work days) • Time delays in the information distribution: 2,344 days less approximately • Less field errors	$19,000	1.3
4 (USA)	2.3 'The tool was somewhat helpful in document and project management'	• Time saving related to RFI process: on average about 40 per cent (2 days) • Time delays related to RFIs: about 326 days less • Time savings: about 365 hours (about 46 work days)	$27,000	22*

Source: Adapted from Sulankivi et al. 2002, p. 50.

Note
* The system used in the US case study was not employed in a comparable way to the other systems, so the monetary benefit achieved was significantly different.

study's results but benefits were found across all three categories (see summary in Table 9.2).

Numerous time and other quantifiable benefits were observed, in particular, time savings as a result of faster distribution of design drawings, allowing work to be conducted to a tighter schedule, and improving individual productivity. Fewer field errors, 'slightly fewer disputes because of documented information exchange' and 'less paper to be filed' were also mentioned (p. 60). Sulankivi *et al.* also identified significant monetary benefits, averaging about US$15,000 across the three European-based case studies. In each instance, the systems more than paid for themselves, returning a monetary saving equivalent to 1.9 times the cost of the ASPs' services, mainly due to savings in communication and document costs, for example: lower mailing costs (average US$2000 per project), less travel, less paper, reduced telephone expenses and more efficient labour (p. 63). The indirect cost savings may well be much greater, as the authors concluded, but are also more difficult to measure:

> In the European case studies, direct cost benefits were approximately double as compared to the operating costs. The indirect cost savings are manifold as compared to the direct cost benefits. However, in practice indirect cost savings are difficult to quantitatively measure.
>
> (Sulankivi *et al.* 2002, p. 91)

9.6   Vendor case studies

In the absence of a wealth of independent empirical research – this section will review some of the limited research available – potential customers have therefore tended to rely on vendor's case studies and surveys in which users give what Orr (2002) describes as 'their (invariably favorable) experiences'. Few of the available vendor case studies offer much by way of measurable benefits, but the author has identified three studies, one published by BIW, one published by the ITCBPP (now the ITCF), and one forming part of the aforementioned ProCE project.

9.6.1   *BIW Technologies*[10]

Kajima Construction Europe (UK) Limited, part of the global Kajima organisation, adopted BIW Information Channel from BIW Technologies during 2000. As one of Asda's partner contractors, Kajima was to carry out a new £10 million superstore development in Wrexham, North Wales, for the retailer. Asda wanted fast-track delivery of the new 45,000 sq. ft superstore, and Kajima set a very tight target for the design and build project.

The Wrexham design phase lead-time was short, and there was enthusiasm from Asda, Kajima and the other suppliers for the concept of using an online system to hold information centrally. BIW Information Channel was implemented for the Wrexham project to good effect. In total, around 30 different companies had access to the system, collectively logging-on some 7,000 times during project delivery, and publishing 400 documents and 1,261 drawings.

Using other innovative techniques and materials during the build phase, the team took just over 13 weeks to complete the superstore – which Asda was able to open a day ahead of schedule. Asda has subseqently used the BIW system on over 30 projects, both new-build schemes and store refurbishments, and to manage standards across its projects.

The reported project benefits included:

- Cost savings of more than £11,000 on printing and £1,000 on postage because client and supplier team members could simply log in to view drawings.
- The project architect reported an 8 per cent increase in efficiency.
- Drawings issue/review/re-issue process reduced from an average of 10 days to just 2 days (best time noted was four hours).
- Around £31,000 was saved on design team fees.
- Fewer site meetings.

9.6.2   BuildOnline[11]

Kier Build began using BuildOnline's project collaboration software during a £59.3 million office development in Hatfield, Hertfordshire, for Arlington Property Developments. The scheme involved design and construction of six, four-storey office buildings around an access deck forecourt with basement parking. Work started in March 2001 and was completed by October 2002 on time and to budget.

Key objectives were established at the start of the scheme to focus on the efficient transfer of documents and the accessibility of accurate drawings. For project collaboration to work successfully the entire project team needed to embrace the use of the software. Therefore, the team set about communicating the message, ensuring everyone wholeheartedly believed in the benefits of using the system. Kier Build recognised that training was vital for successful implementation, and arranged for this to be provided through half-day workshops and web-conferencing sessions run by BuildOnline's customer services team. During the project, support was provided by a direct line customer services help desk, an email query service and an online help facility.

The main issue that Kier Build faced was effective adoption of the new software and way of working. Each of Kier Build's supply chain members had different levels of IT infrastructure and competency. To manage this, a 'project champion' was appointed to ensure successful adoption of the software across the project team. All team members received training, at the outset and throughout the project.

Adopting the BuildOnline system yielded an overall improvement in productivity costs for the team. These included:

- Cost reductions of £40,000, generated from the reduction in copying, postage, couriers and telephone calls.
- Time saving of 2,500 work-hours in administration time for document distribution.
- The drawing review process being reduced by seven working days.
- A complete audit trail of all documentation.
- The client being able to monitor document movement.
- Increased trust and transparency across the project.
- A complete archive of project information.

9.6.3 Sarcophagus

The third project studied by Sulankivi *et al.* (2002) was a £16 million retail scheme in south-east England, the 6900 m^2 store being part of a larger nine hectare development. All the main parties were located within a 300 km radius of the project location; design started in mid-summer 2000, with steel erection begun the following spring and completion and hand-over in autumn 2001. Excluding site preparation and adjoining development work, store construction was scheduled to take 17 weeks.

To facilitate electronic publishing, dissemination and archiving of project documents during the construction phase, the team used 'the-project', a construction collaboration application from Sarcophagus. Documents included drawings, specifications, component schedules (e.g. finishes, reinforcement), progress reports, meeting minutes, approved changes, anticipated final expenditures, health and safety (CDM) documents and risk assessments.

In most cases, the Sarcophagus system replaced posting, emailing, or faxing drawings and other documents to different parties, and also served as a centralised project archive for documents, contact information and site photos. The owner, design team, construction manager, main contractor and several sub-contractors all used the system.

Interestingly, the client established and enforced a directive that all designs be drawn at a scale appropriate for reproduction on A3-size paper whenever allowable. This eliminated all A0- or A1-sized drawings except for some specialist inputs, mainly from mechanical and electrical engineers, and significantly influenced the way the CE-environment was used. Because designs were on A3-sized paper, nearly all were distributed electronically between the main parties, and each was responsible for printing document copies to fulfill their own needs.

As part of the ProCE research project, benefits measurements were carried out as described in Section 9.5, with 17 questionnaires received and analysed, including responses from the main designers, sub-consultants, main contractor, sub-contractors and inspectors. As summarised in Table 9.2 earlier, the researchers believed the Sarcophagus system cut delays in distribution of information, saved about 20 work days and cut the number of field errors. The target to reduce paper was reached 'to some extent'. The total monetary saving was estimated at £12,000 (US$19,000), more than outweighing the cost of the system. The designers also predicted operation and maintenance benefits: project information is maintained in a useable form and remains available to future users on the service provider's server, and a full electronic archive is created for the client.

9.7 Calculating an ROI

Construction collaboration technology vendors are often asked to provide ROI examples that a customer might then use to help justify its own investment in that technology. This may seem a simple request but given the sheer range of potential costs and benefits already outlined, it will be clear that a comprehensive ROI calculation will seldom be possible. While some US providers (notably Constructware and Buzzsaw) have developed rudimentary ROI calculators, as already mentioned (Section 9.1), most UK providers tend to rely on anecdotal evidence of the savings to be achieved.

9.7.1   Tangible savings

Even if we focus on just the tangible or hard cost savings, ROI calculations can still be very complex. The range of variables will clearly vary from project to project and will also vary according to the number and type of organisations involved, their normal working practices, existing familiarity with and use of IT in general and collaboration applications in particular, etc. However, at its simplest and most tangible level, the initial calculation of potential savings may take the following $A \times B \times C$ form:

$$\text{Total saving} = (\text{number of items}) \times (\text{cost of production of items})$$
$$\times (\text{number of recipients of items})$$

Assuming construction collaboration technology will largely replace conventional production and distribution of design drawings, specifications and other hard copy documents, this might involve consideration of:

- total number of drawings to be produced;
- total number of non-drawings to be produced;
- average costs of printing and copying documents (mainly paper and consumables);
- average man-hour costs for folding/preparing hard copy documents for distribution;
- average man-hour costs for general drawing and document management;
- average costs of posting or – if urgent – courier delivery of hard copy documents;
- total number and average costs of faxes;
- average costs of storing hard copy documents;
- allowance for costs of reprinting, preparing and posting packages lost in transit;
- total number of individuals on issue lists for above.

9.7.2   Tangible costs

First and foremost, the costs the construction collaboration technology will need to be taken into consideration. Assuming this is rented on an ASP basis[12] regardless of the number of users, number of documents or volume of storage, the factors to consider might include:

- total cost of renting technology ($\times$ months at £y/month);
- costs of implementation consultancy and set-up;
- costs of training (including 'down-time' of end-users while attending training, etc.).

Some supply chain members might need to print out some drawings or documents locally, and this may be reflected back to the customer in terms of higher fees or tenders:

- total number of drawings to be produced;
- total number of non-drawings to be produced;
- average costs of printing and copying documents (mainly consumables);
- allowance, where necessary, for acquisition of new printing or plotting equipment;
- minus: savings from printing out drawings or drawing sections at smaller sizes.

Similarly, some supply chain members might need to acquire new IT equipment or upgrade their existing infrastructure in order to access the service effectively (again, these expenses may be reflected back to the customer in terms of higher fees or tenders – though such improvements cannot reasonably be associated solely with a customer's project):

- additional hardware, software or telecommunication upgrade costs incurred;
- additional internet connection costs incurred (more frequent access or higher data volume charges).

9.7.3   Some real ROI examples

It might be helpful to quantify the potential figures involved on some of these areas. In addition to the ProCE case studies (Sulankivi et al. 2002; see Section 9.5.1), in 2000, one of the earliest UK projects (a 15-month, £23 million scheme) to use an internet-based collaboration system (from BIW) was the subject of a confidential study by a researcher from the University of Salford. Adopting the simple $A \times B \times C$ calculation mentioned above, the researcher identified postage savings in excess of £6,000 and printing savings of over £41,000. While the lead consultants undoubtedly made significant savings in printing and distributing drawings and other documents, these needed to be balanced against claimed additional printing costs further down the supply chain, but these only amounted to around £2,600. Accessing the collaboration system was estimated to have cost a total of around £3,600 (this figure was not adjusted to account for pre-existing internet connections). Assuming the collaboration system in question cost, say, £1,000/month, and there were, say, implementation and training costs of £5,000, the system costs would have been £20,000. The overall tangible saving on communications on this project was over £47,000; taking away the system costs and the additional supply chain expenses, the net tangible saving was still some £21,000, a ROI of almost 80 per cent (see Table 9.3).

Howard (2002) outlines another example provided by UK contractor Balfour Beatty who had been using BuildOnline's system for a £20 million project. Savings of about £24,000 were achieved on one work package; estimates on 10 different types of work package predicted savings of £144,565 against costs of £55,208, a net saving of £89,357, a potential ROI of about 160 per cent. Howard also reviewed use of a Danish system, Byggeweb, but the quantifiable benefits had yet to exceed the costs,

Table 9.3 The UK construction project – a real life ROI example

Costs	£	Savings	£
ASP subscription (15 months @ £1,000)	15,000	Postage	6,000
One-off ASP set-up, training, etc.	5,000	Printing	41,000
Internet access costs	3,600		
Additional sub-contractor expenses	2,600		
Total cost	26,200	Total savings	47,000
Net saving (savings – costs)			20,800
ROI (net saving/total cost)			79.4%

probably because as this was an experimental project in which email and post continued to be used in parallel.

9.7.4 Developing the ROI model further

Moving beyond a somewhat conservative focus on the tangible communications savings, ROI calculations might take into consideration a longer list of hard and soft benefits (see Section 9.2), any additional hard and soft costs incurred (see Section 9.3), and – particularly if the collaboration technology is offered as part of a package of services by a consultant or contractor – any new fee-earning or revenue-generating potential (see Section 8.3.7). Even if we just focus on the ROI to an industry client, it soon becomes clear that the potential benefits will greatly outweigh the potential costs even on just a single project – see Table 9.4.

Various factors will influence the model illustrated in Table 9.4, including:

- *Organisation type* The above model is based on supporting an AEC industry client, and would need to be adapted where the customer was, say, a construction manager or contractor.
- *Project load* Will the collaboration application be used on a single project, on a staggered series of projects, or a multi-project programme where lots of schemes will be undertaken simultaneously?
- *Pay-as-you-go use or corporate deal?* As mentioned in Chapter 8, vendors may offer considerable pricing benefits where a customer enters into a long-term strategic relationship.
- *Existing IT support* Will the application replace an existing in-house system? If so, the potential savings can be much greater due to the efficiency of ASP service delivery over an internally managed solution.
- *Contractual issues* Will the project team be working collaboratively or using the system to maintain traditional relationships? How far down the supply chain will use extend? This is quite important: the costs and benefits may not be shared equally across all team members. Sub-contractors down the supply chain may feel they have little to gain, while there may be considerable cost savings for the lead designers and project manager (some projects have successfully employed pain/gain-sharing arrangements to share the costs and benefits of using collaboration technology across the team).
- *Extranet experience* The performance of a project team will also vary according to its members' previous experience (if any) of using such technologies.

9.8 Chapter summary

This chapter – the penultimate chapter in this book – has described the benefits that the vendors claim will result from AEC industry clients and their project teams using construction collaboration technologies, and, through some case studies and research studies, has reviewed the evidence to substantiate those claims. While it is difficult to provide conclusive evidence to support and quantify all the claims – particularly those relating to intangible benefits – there is enough evidence to suggest that these claims

Table 9.4 The UK construction project – an ASP customer ROI model (in dollars)

Costs	Savings
Hard costs	*Hard savings*
ASP hosting subscription (£/month)	Postage
ASP implementation and set-up fees	Couriers
ASP training fees	Printing
ASP provision of archives	Storage/archiving across multiple sites
Additional team member expenses,	Travel and meeting cost savings,
for example	for example:
• additional internet access, IT costs	• fewer, but more effective meetings
• additional printing costs	Telephone/fax savings
• training time	
Total hard costs	Total hard savings
Short-term soft costs*	*Short-term soft savings*
Offline duplication of information	Time saved (e.g. lower consultant fees)
Needless drawing revisions	• Faster mobilisation
Superfluous comments	• Faster access to information
	• Less time searching
	• Less re-keying, scanning
	• Less administration/form-filling
	• Fewer but faster drawing revisions
	• Faster hand-over documentation
	• Faster RFIs and other processes
	• etc.
	Better IT compatibility/IT savings
	More collaboration
	Fewer changes, mistakes/less rework
	Earlier completion (earlier revenue)
	Fewer claims or legal disputes
Long-term soft costs	*Long-term soft savings*
	For example:
	• better designed and built facility
	• more efficient O&M/FM
	• reuse of data on other projects
	• better management of standards
Total costs	Total savings
Net saving (savings − costs)	
ROI (net saving/total cost × 100)	

Note
* It is assumed that such short-term costs will eventually disappear as users become more confident/expert.

are not just 'hot air'. Certainly, end-user enthusiasm for the systems does suggest that they solve more problems than they create. However, it is also clear that project teams need to do more to objectively measure the impacts on their projects of using the technology. And in an effort to help businesses think about the key variables, the chapter ended by proposing an outline that could be used to start calculating an ROI.

Since defining the technology in Chapter 1 and then describing the development of the UK construction collaboration technology market in Chapter 2, this book has concentrated on helping the reader learn and understand more about the different variables to be taken into account in selecting and adopting a collaboration system. Chapters 3–9 gave a detailed account of all the key issues, but it will be clear that construction collaboration technologies are continually developing and there is still considerable scope for change within the market for them. This book therefore concludes with a look to the future. Reflecting the continued evolution of the AEC industry, its IT tools and the telecommunications at its disposal, Chapter 10 speculates about some of the potential developments that may influence the future use of collaboration systems.

Where next for construction collaboration technologies?

This chapter:

- draws together several themes and trends discussed in previous chapters and speculates about their possible future convergence;
- underlines that, notwithstanding rapid developments in IT and telecommunications, the fundamental drive towards greater collaboration will depend on changing industry, organisational and individual attitudes and working practices;
- discusses the critical tests – reliability, ease-of-use and cost-competitiveness – that construction collaboration technologies must overcome to encourage widespread adoption.

Much has been written over the past few years about the 'internet revolution', but so far as the AEC industry in the United Kingdom is concerned, it is perhaps better to describe the process more as 'evolution' (after the dot.com boom and bust, the notion of 'survival of the fittest' is probably quite apt). As we have seen, the AEC sector's conservatism has been a factor in the slow adoption and use of new technologies, and the gradual increase in popularity of construction collaboration technologies has occurred against a background of industry culture changes and of IT advances, particularly since the mid to late 1990s. These changes have not stopped. The industry, the IT tools and the telecommunications at its disposal continue to evolve.

Some developments (cultural, corporate and technological) are already under way, and there will doubtless be a steady flow of new ones in the months and years ahead. This chapter will expand on some of the trends discussed in previous chapters:

- the development of a more collaborative working culture (Chapter 2);
- market rationalisation among the collaboration providers (Chapter 3);
- remote hosting of construction collaboration technologies (Chapter 4);
- the emergence of new technological capabilities (Chapter 5);
- new connectivity possibilities (Chapter 6);
- new regulatory pressures (Chapter 7);
- changing working patterns (Chapter 8).

A detailed consideration of every scenario is beyond the scope of this book, but in the early twenty-first century there are numerous disparate trends that may, or indeed may not, affect how collaboration technology continues to develop. This chapter will discuss some of the most salient trends and attempt to identify some of the common themes that may emerge.

10.1 Trends in collaborative working

As described in Chapter 2, organisations and individuals within the AEC sector are still developing their own approaches to the notion of working more collaboratively. Whether described as 'partnering', 'integrated working' or 'collaborative working', more progressive methodologies have yet to become the predominant way of delivering construction projects (as mentioned at the end of Chapter 1, industry estimates in late 2004 suggested only around a third of all projects embraced more progressive approaches). The slow adoption of collaborative working probably has much to do with the fragmented and often deeply entrenched approaches of many organisations within the industry, but the increased emphasis placed (first by Latham, then Egan) on satisfying the client and on managing whole life costs has begun to change the environment.

From their position as paymasters at the top of the project 'food chain', clients are clearly able to profoundly influence what they want from their project teams, and if the efforts of leading client organisations (not least the many UK government bodies, both central and local) continue then the predominance of potentially adversarial forms of project delivery may well diminish further. As AEC industry clients and their project teams become more enthusiastic about collaborative working, they will increasingly switch from systems – Chapter 1 described them as 'construction communication technologies' – that mimic traditional project information processes and controls (allowing some team members to electronically restrict access by some team members to certain types of information) to technologies that encourage collaboration, openness and trust.

Moreover, clients are increasingly also being urged to look beyond the initial stages of designing and building a new asset, and consider how that asset might best be managed throughout its operational life. For example, a paper by the Royal Academy of Engineering (Evans *et al.* 1998) suggested that buildings designed for the accommodation of people generating wealth, or people providing a service, must create an environment where people will give their best:

> The cost of ownership and maintenance of a building is typically about three per cent of the overall cost of the people working in the building. A useful guide for the whole life cost of operating and owning commercial office buildings is illustrated by the following ratios:
>
> | | |
> |---|---|
> | Construction Cost | 1 |
> | Maintenance and Building Operating Costs | 5 |
> | Business Operating Costs | 200 |
>
> Similar ratios might well apply in other types of building. There is a good deal of evidence that the building itself, if properly designed and managed, can lead to significant improvements in productivity.
>
> (p. 5)

Some forward-thinking UK practitioners enthusiastically endorse such thinking. Richard Saxon (see Crane and Saxon 2003), for example, argues:

> The leverage between thought input at the start and performance benefit generated therefore runs to a ratio of potentially 1:2500 upwards. Thoughtless downward pressure on design time is therefore more than a false economy; it is economic non-sense and at the root of our national dissatisfaction with the construction industry and with our built environment generally.
>
> (p. 60)

With post-construction building-related costs estimated to be five times greater than the cost of designing and constructing the building, a strong business case can clearly be made for capturing as-built information for future FM purposes. Already, an increasing number of public sector facilities are being delivered through prime contracting, PFI and PPP projects where funding, design, construction and long-term operation and maintenance responsibilities are concentrated in the hands of a single 'special purpose vehicle' or consortium. Such projects have encouraged the adoption of a 'whole life cost' approach to the procurement of assets such as schools and hospitals, and have been enthusiastic about the use of construction collaboration technologies to manage the huge volumes of documentation that are compiled from initial bid stages, through project delivery, to operation and maintenance.

The United Kingdom is not alone in wanting more coherent, long-term approaches to project delivery. Similar views are also expressed in the United States. For example, a working party of the Construction Users Roundtable (CURT) published a pamphlet in August 2004 which echoed for its US readers many points also made repeatedly in the United Kingdom. The vital, leading role of the ultimate client is emphasised (e.g. 'Owners, as the integrating influence in the building process, must engage in and demand that collaborative teams openly share information and use appropriate technology' – p. 3) before the CURT's vision includes an explicit 'whole life' role for IT that could just as easily be applied to the AEC sector in the United Kingdom:

> Digital information created by the collaborative team flows throughout the lifecycle of the building project, emanating from the building information model that has virtually constructed the building before construction commences and supports its operation throughout its life. The information permeates every aspect of the building's lifecycle.
>
> (p. 12)

The extent to which clients can effectively lead the way towards more long-term integrated or collaborative relationships will depend on how much work they need to procure. The fascinating comparison by Green *et al.* (2004) between the AEC and aerospace sectors concludes on a somewhat pessimistic note. The authors suggest that regular workflow is vital before firms can begin to compete on the basis of innovation rather than short-term cost efficiency, while such a shift depends crucially on a willingness to invest in new skills and new methods of human resource management.

However:

> clients must provide continuity of work if they are to promote any lasting change. Such developments are likely to be limited to an elite group of firms that serve the needs of major repeat clients. Notwithstanding the unlikely occurrence of widespread 'cultural change', the prospects for a general shift to collaborative working are limited by the structural characteristics of the sector.
>
> (p. 84)

Clearly, only a few prolific clients will be able to offer such continuity of workload, or there may be a need for an over-arching programme to bring together large numbers of individual clients. For example, the UK's Partnerships for Schools programme, Building Schools for the Future, is targeted at 3,500 secondary schools, from every local authority, over a 15-year period.

10.2   Market rationalisation

In the AEC industry sector, which sees a steady stream of business mergers, acquisitions and liquidations, the emergence of construction collaboration technology providers during the 2000 dot.com boom quickly prompted numerous warnings that buyers needed to be wary of what many commentators saw as an inevitable market consolidation. Such warnings were unlikely to affect one-off clients intent on streamlining the delivery of a single project, but they doubtless had an effect on more prolific or regular industry clients, plus many contractors and consultants. Many such organisations adopted a 'wait-and-see' approach, and by 2004 even the most optimistic surveys (e.g. IT Construction Forum 2004) were still showing that take-up of the technology had yet to reach more than half of all the UK projects. As a result, a substantial market – both in the United Kingdom and in many other countries – remained as yet untapped. While the leading vendors had sought to win over many of the major industry clients and leading contractors and consultants, there probably remained enough prospects for the leading vendors not to encounter particularly intense, 'cut-throat' competition. However, as take-up extends, one might expect the battle to win and then retain new customers to become more competitive.

By late 2004, of the providers highlighted in Chapter 3, probably only 10 have both established a significant UK market awareness and built a viable UK customer/ user base in construction collaboration (in alphabetical order):

* 4Projects
* Aconex
* Asite
* BIW Technologies
* BuildOnline
* Business Collaborator
* Buzzsaw
* Cadweb
* Causeway
* Sarcophagus.

Of these, the 'big boys' in terms of the size of their user base have tended to be 4Projects, BIW, BuildOnline and Business Collaborator, and while it would be unwise to completely discount all of the others (e.g. Asite is backed by a wealthy property tycoon, Aconex is looking to build on a strong Australian track record, while Buzzsaw and Projectwise stand to benefit from the large Autodesk and Bentley footprints in the AEC design market), some may well be content to grow slowly while maintaining efficient services for their existing customers and end-users. Encouraged by their experiences to date in the United Kingdom, one might also expect the 'big boys' to extend their reach outside the United Kingdom, although it is likely that any dramatic expansion or merger and acquisition activity would require additional funding, perhaps from venture capitalists or through a market flotation.[1]

There is also still considerable space in the market for other collaboration tools (e.g. the Lotus IBM QuickPlace package, Microsoft SharePoint, or the peer-to-peer (P2P) based Groove Virtual Office) or the increasingly collaborative features of more generic software applications (e.g. Microsoft's Office tools, Adobe Acrobat) to be used to support the smaller projects often ignored by some of the above, more specialist names.

Within the AEC collaboration market, much will depend on how each business responds to the different economic, competitive and technological challenges and opportunities that lie ahead. For example, increased rates of technology adoption coupled with migration towards a small number of vendors will test the capabilities (both human and technological) of the chosen vendors to respond. An ASP vendor will need to be able to anticipate and manage its new workload so that its services do not become becomes slow or erratic, while a vendor of customer-hosted solutions may need to expand its implementation and customer support resources.

However, consider what might happen if two ASP-based construction collaboration technology businesses were to merge. On the face of it, one business would potentially be acquiring a whole new mass of customers and end-users, and the combined figures would probably be very impressive (assuming, of course, that the new management team was able to keep 'churn' to an acceptable level). However, the managers of the newly merged businesses would face some difficult choices. Do they continue to invest human and technological resources to market, implement and support two different products on two different hosting environments?[2] Do they seek to migrate the users of one system to the other (risking a backlash from users who preferred the discontinued product, and potentially affecting the scalability, speed and reliability of their hosting environment for users of the retained system)? If the discontinued system was cheaper, will the new alternative be offered at the same original (but unviable) rate, or will the management accept that customers may decide to seek an alternative supplier?

A factor in any market rationalization will be the ease with which data can be transferred between different vendors' systems. As mentioned in Chapter 3, some customers wanted the reassurance that they could switch quickly to another system in the event that their chosen provider ceased trading or the commercial relationship needed to be terminated. With many end-users being mandated to use particular systems by their customer, some users – particularly those engaged simultaneously on different projects each using different systems – wanted to cut re-training requirements and be able to use one familiar interface to access information regardless of which vendor provided the underlying system hosting the data. Responding to these needs, several of the leading UK providers of construction collaboration technologies sought to address customer and user concerns about the perceived lack of industry standards for these types of systems. One of the motives for the 2003 establishment of the Network of Construction Collaboration Technology Providers (NCCTP) was to devise interoperability standards to which its member providers could comply. In late 2004, the XML-based data exchange standard, allowing information to be exported from one provider's system and imported into another's, was being tested by NCCTP members (4Projects, BIW, BuildOnline, Business Collaborator, Cadweb, Causeway and Sarcophagus).

The development of a standard to allow interoperability between the different collaboration platforms and data held by different hosts has to be regarded as a more long-term project, and raises some interesting possibilities. For example, would the current status quo continue? Would end-users stick with the interface technologies (i.e. software interface and file viewer) with which they are most familiar, and continue to use them to access project data held on all compatible collaboration platforms? And would each provider then continue to market their interface, underlying platform and hosting capabilities? Or would there be a gradual change as end-users progressively migrated to whichever interface combination(s) became accepted as the most attractive, functional, intuitive or easy to use? Could there be a time when some providers give up providing their own interface and focus on providing the underlying collaboration application and/or hosting facilities, perhaps working in a symbiotic relationship with the firm(s) which now specialise solely on being leader(s) in interface technologies?

10.3   The continued rise of the ASP?

What has particularly distinguished the emergence of web-based construction collaboration technologies from the adoption of other IT tools in the AEC sector has been the widespread adoption by vendors – and the equally widespread acceptance by customers – of the ASP model of software delivery and of subscription-based licensing arrangements (see Chapters 3 and 4).

It is perhaps too early to suggest that we have begun to see the end of traditional licensing models – with software bought per seat and for a given period of time and in relation to a particular version – but there are signs that buyers are becoming resistant to what they see as an inflexible system replete with tedious upgrade processes and onerous maintenance fees. Buying web-delivered software by paying a monthly subscription is already becoming popular in several other sectors. For example, in addition to construction collaboration applications, the subscription model has been successfully applied to several mainstream applications, including, among other things, customer relationship management, enterprise resource planning and accounting. As a result, ASP companies such as salesforce.com, RightNow and NetSuite have become regarded as serious competitors – particularly at the lucrative SME end of the market – to the likes of Siebel and SAP, but the software majors such as Microsoft and Oracle may still prove resistant or slow to change, having already made billions from traditional licensing arrangements. With the UK construction industry being so fragmented and dominated by SMEs, it is likely that the subscription model will become increasingly attractive as awareness grows.

Another potential consequence of the emergence of the ASP model could be the demise of traditional IT functions or corporate IT departments.

> In time,…companies will be no more likely to run their own computer systems than they would generate their own electricity. And the traditional software program in a box could one day look as quaint as the water pump in the yard.
>
> (Taylor 1999)

10.4 New collaboration technologies

The pace of technological development shows no sign of slowing down, and some of the many innovations currently being discussed and even tested may well become run-of-the-mill by the end of the decade. For this book, four key areas of innovation have been identified as having a potential impact on the UK construction collaboration technology market:

* integration with back-office and other systems;
* real-time collaboration;
* building information modelling (BIM);
* radio frequency identification devices (RFIDs).

10.4.1 Integration with back-office and other systems

The emergence of the ASP model has prompted lots of discussion about how web-based services can be integrated to work with other systems. In the past, integration between different systems was enabled by special software known as middleware, perhaps by reference to the electronic data interchange (EDI) standard, but in recent years, various inter-related initiatives specifically focused on web services have begun to emerge. It is, however, a complex and constantly evolving picture (made no easier by the confusing array of abbreviations – jargon-haters should skip the next paragraph).

XML is increasingly being used as a way to standardise data formats and exchange data, and forms the basis of Web Services Description Language (WSDL). WSDL is used to describe the services a business offers and provides a means for individuals and other businesses to access those services electronically. WSDL is also the key language used by Universal Description, Discovery and Integration (UDDI), an XML-based registry, which enables businesses to list themselves and their services on the internet (WSDL supersedes Microsoft's Simple Object Access Protocol (SOAP) and IBM's Network Accessible Service Specification Language (NASSL), used to help programs running on different operating systems communicate with each other).

The key will be to agree the standards necessary to exchange data about the building and its components between the different applications employed within a project team. An international standards-setting body, the International Alliance for Interoperability (IAI) has been particularly active in this field, developing the Industry Foundation Class (IFC),[3] a data representation standard and file format that defines architectural and constructional CAD graphic data as three-dimensional (3D) objects (with data about dimensions, manufacture, relationships to neighbouring components, thermal and acoustic performance properties, maintenance needs, etc.).

Almost in parallel, Bentley Systems headed the complementary development of the aecXML standard, designed for non-graphic data. Between the two, it is hoped that a set of vendor-neutral standards will emerge, allowing easy exchange of object data – both the rich descriptive content of the IFC and the small packets of information needed for business transactions that are managed using the aecXML standard – between CAD, estimating, quantity surveying, project management and other systems.

Do not think this is an issue that should only matter to IT specialists. It is sobering to reflect that one major American study, from the National Institute of Standards and Technology (NIST 2004) calculated that poor interoperability[4] between CAD and other systems cost the US construction industry a massive $15.8 billion a year, or – at a conservative estimate – between one and two per cent of the industry's annual turnover, with owners and operators bearing about two-thirds of the estimated costs.

Costs include manual re-keying of data, duplication of business processes, and substantial reliance on paper-based systems for production, distribution and storage of information. Moreover, there is little reason to suggest that the UK experience will be substantially different; on an AEC industry turnover of some £83 billion, similar lack of integration between data and systems is likely to be costing the UK construction industry somewhere between £800 million and £1.6 billion per annum. Nor is the issue one that can only be resolved by the technology vendors. As we have already discussed, continued reliance on paper-based systems, inconsistent levels of technology adoption up and down supply chains, and industry fragmentation are all factors that affect the AEC sector as a whole.

There are, however, already signs that greater integration is on the way, with applications converging from both generalist and specialist software providers. Generic software applications, such as Microsoft's Office tools, increasingly incorporate functionality designed to help users share information, Microsoft's SharePoint is explicitly marketed as helping teams, regardless of their profession or industry, to share documents and workspaces. From the AEC perspective, at a basic level, some construction collaboration technology providers have already been using XML-based approaches to achieve integration between some users' locally hosted legacy applications and their online project environments so that information does not need to be managed twice.[5] And, at a more strategic and far-reaching level, greater use of BIM (see Section 10.4.3) could provide the key to improved integration between AEC firms. Greater integration may also come via new developments in computer operating systems. For example, Microsoft's next major operating system release will feature enhanced collaborative functionality; codenamed 'Longhorn', industry gossip in late 2004 suggested it was scheduled for launch around May 2006.

Such integration initiatives are likely to extend to many other applications. Depending on their role or responsibilities, users may want to exchange information with estimating, project accounting and/or corporate financial reporting systems. They may need to view data held in corporate intranets, project scheduling or customer relationship management systems. Greater focus on 'whole life costs' may prompt exchange of data with FM and space planning applications, etc. Already, some collaboration application are allowing users to retrieve, view, sort and manipulate data from multiple projects, and the time may soon come when these applications can retrieve data held on multiple systems – even those hosted by competitors. Some of these applications may be web-based, others may be conventional client/server architecture, but the overall effect of greater integration may well be to blur the edges of collaboration applications and to obscure the inputs of different software developers and data hosts. In short, collaboration technologies may become just 'part of the plumbing'. Through his or her desktop, the end-user may be able to seamlessly access and manipulate relevant information drawn from a whole variety of integrated, interoperable sources.

10.4.2 Real-time collaboration

As this suggests, construction collaboration technologies as we currently conceive them may only be the start. The next few years may well see collaboration move to a new level: real-time exchange of ideas and information.

Real-time or synchronous collaboration means using technology to communicate with fellow team members as if they were in the same room, even though they may be located on the other side of the world. Chapter 1 highlighted the difference between synchronous and asynchronous collaboration, and, to date, most of the vendors of construction collaboration applications have focused on the latter: supporting activities where collaboration involves a succession of interactions occurring over a period of time. For example, a conventional drawing approval process might involve several steps – issue of the drawing, receipt of feedback/comments, issue of new revision of drawing, final feedback/comments, final revision, approval – all documented by successive exchanges of paper.

Synchronous collaboration takes place when participants review and discuss issues in real time. Such forms of collaboration can be very productive as they have more 'emotional bandwidth' than, say, email exchanges, helping build trust and rapport between team members. Traditional real-time collaboration might have involved face-to-face meetings, telephone conversations, perhaps – combining elements of both – a tele-conference. However, advances in internet technologies now mean that construction collaboration vendors can – and in all probability will – add new capabilities to their systems to further accelerate and improve the quality of decision-making processes. Many of these synchronous technologies are already available, but take-up has, to date, varied widely both across different industry sectors, between different organisations and within organisations. Such tools include:

- group chat
- 'presence awareness' technology
- whiteboard collaboration
- application sharing
- desktop or screen sharing
- voice over IP (VoIP)
- video and audio web conferencing tools.

Most instant messages involve exchanges of short text notes, although some systems offer file-sharing and allow users to talk to and see one another through the use of cameras, microphones and speakers or headphones. While there have been issues relating to security, archiving and manageability of non-enterprise IM solutions, IM and web conferencing in particular have become extensively used as stand-alone tools within many organisations to facilitate person-to-person communication, and are gradually being applied to external situations where staff might need to collaborate in real time with customers, partners and suppliers. There remain numerous challenges; for example:

- Achieving greater interoperability between the various technologies (e.g. a user of Yahoo's IM solution could not communicate with users of other popular consumer IM tools from AOL and Microsoft).
- Achieving consistency of performance via the internet (this could well change as the mobile connectivity challenges are overcome).
- Overcoming cultural issues relating to new ways of working (as we saw in Chapter 8, training is essential for breaking down barriers and setting expectations; guidelines, such as 'what constitutes an urgent message', need to be established, but should not be too restrictive in case they stifle team dynamics).

However, some of the big-name IT vendors (e.g. IBM, Microsoft, Oracle, Siemens, Sun Microsystems) are promising and in some instances delivering tools that embrace standards-based technologies such as XML web services and VoIP. This will provide much-needed flexibility to allow collaborative components to be sewn together and embedded in various applications, launched from numerous communication tools, and consumed by just about any device.

Users would, for example, be able to log in to a construction collaboration system, access documents and then initiate ad hoc real-time collaboration with fellow team members – assuming, of course, that these members have indicated their online availability, and that the context of the required collaboration is clear (e.g. if the discussion concerns a drawing comment, then that drawing is immediately accessible on each person's desktop). A drawing approval process might be completed much more quickly by allowing feedback to be delivered in real time. For instance, a designer might call, schedule and host a 'net meeting' at which participants each sit in front of their own computer screen. Suggested changes could be sketched out 'live' and discussed simultaneously, with each conclusion highlighted via a comment and/or mark-up indicating the required amendment. A session transcript (or even audio or video files), detailing the participants and the issues, documents and drawings discussed, might also be saved for audit trail purposes.

10.4.3   Building information modelling (BIM)

The idea of using intelligent 3D models of buildings instead of 2D drawings to develop design ideas and guide construction has been around since at least the arrival of CAD solutions in the 1980s, but to date only a small minority of AEC industry clients, designers, contractors and suppliers have begun to translate the vision of a single, shared electronic model, based on building data rather than geometric data, into reality.

There has been much greater enthusiasm for BIM in other industry sectors (e.g. automotive, aerospace and other mechanical engineering markets), but what works when integrated teams of designers and manufacturing experts collaborate in a large-scale, mass-production, factory-based environment may not be appropriate for fragmented and geographically dispersed AEC teams intent on delivering a complex, one-off project in a specific, outdoor location to a tight budget or programme. This does not mean that the opportunity has been completely ignored. Indeed, many building components and systems (from cladding and steelwork to building services, doors to lifts and escalators) are increasingly designed using the latest computer aided design and manufacturing (CADCAM) applications and then manufactured or fabricated off-site. Architects and, in particular, structural engineers have begun to use 3D tools to refine their designs and identify potential problems before construction starts on site (e.g. giving their customers virtual reality (VR) walk- or fly-throughs, modelling different lighting schemes, using them for 'clash detection', etc.); a 2004 survey of CAD managers (Davies 2004) found a structural engineer was four times more likely to be using BIM than an architect (p. 9).[6] And the building model could also be used to test, evaluate and refine proposed construction or maintenance processes. A time-dependent (4D) model could be used, for instance, to show the sequence of activities being undertaken at different stages so that the team can highlight potential conflicts between different activities or trades, or spot potential health and safety issues. BIM could also be used, in due course, to model other dimensions, for example, cost or sustainability. But we have yet to see widespread sharing of intelligent building models to help collate and coordinate the inputs of every AEC project team member. Why?

In some respects, BIM faces similar cultural, organisational and technological hurdles to those that have prevented or delayed the introduction of collaboration technologies into the project team. Whether one is talking about conventional 2D working, new-fangled 3D, or 'nD' covering other dimensions, as was stressed in Chapter 8, owners (particularly influential as they ultimately fund facilities and must bear the cost of their future operation and maintenance), their designers, contractors and supply chains need to be committed to the idea of collaborating. They also need to be motivated to collaborate. And they need the appropriate knowledge, skills, tools and infrastructure to help them collaborate electronically.

Finally, and perhaps most pertinently, the costs of many BIM applications remain somewhat prohibitive;[7] it requires a collective willingness by all team members to invest in and use the technologies, and the applications used can place heavy demands on hardware and telecommunications networks. Nonetheless, assuming that (a) AEC clients in particular, but also their consultants and contractors, increasingly begin to demand BIM approaches to the delivery of capital assets, (b) that the BIM application developers can price their packages at a level that encourages wide adoption and use, and (c) that construction collaboration technology vendors can respond to the challenges of sharing building model information in the same way that they have responded to managing conventional 2D CAD files, then the wider use of BIM within the collaboration environment may not be far away.

10.4.4 Radio frequency identification devices (RFIDs)

As well as managing information about building components using IFC-based building models, the components themselves could carry the same data, using miniaturised transponders or RFIDs. Unlike barcodes used to label categories of items, these machine-readable auto-data collection devices give each tagged object a unique identity, and are readable by RF equipment without contact or line of sight. Being able to withstand harsh environments, RFIDs have already been used extensively for identification, tracking and security purposes. Attached to or even implanted within a (non-metallic) building component during manufacture, it would in theory be possible to use the substantial storage capacity of each tag to track the arrival of that component on site, monitor its installation and commissioning, and then, when necessary, interrogate it in situ for FM purposes (each item might carry data about its specification, maintenance requirements, service history, etc.). However, as with some of the other technologies discussed, barriers remain. Again, there are no agreed hardware standards, meaning that some manufacturers' RFIDs will not 'talk' to equipment made by other manufacturers. Such proprietary approaches mean RFID costs remain relatively high, though this is likely to change once standards become established.

10.5   Increasingly mobile connectivity

The rapid pace of new developments in telecommunications has helped accelerate the adoption and use of web-based construction collaboration applications. So far (see Chapter 6), this book has tended to concentrate on the growth in the number of broadband subscribers in the UK market, but we have also witnessed enormous growth in mobile connectivity. This is nothing particularly new. As organisations increasingly began to embrace mobile working during the 1990s, they sought to provide their employees with mobile technologies to help them keep in touch. Equipped with, say, a laptop computer, a mobile telephone and the appropriate cable and data card, users could make dial-up connections to the internet, enabling them to manage their email, access corporate networks and view information on the web, etc. – albeit at a slow rate of between 9.6 and 14 Kbps. However, new generations of mobile connectivity options have begun to emerge:

- wireless local-area networks (WLANs)
- General Packet Radio Services (GPRS)
- 3G – 'Third Generation' – communications
- Broadband wireless applications (BWI).

WLANs, also known as WiFi,[8] offer an alternative way for users to log-in to a corporate network or internet gateway. Instead of physically connecting via an Ethernet cable, users can gain access via fixed radio transceivers or base stations (a typical single 802.11 b standard access point can support 15–20 users, has a range of up to 300 m outdoors, and can provide up to 11 Mbps of bandwidth); end-users have antennae and adaptors that either plug in to, or – increasingly – are installed as standard components within their computers. Moreover, users are no longer confined to sitting at a desk next an Ethernet connection point but can move around retaining connectivity so long as they remain within range. WLANs can be used to extend network accessibility to areas where cabling might not be cost-effective or practicable, and are becoming increasingly common outside the office. For example, some airports, railway stations, cafes, hotels and conference centres are creating internet connection 'hotspots' that can be accessed by travellers, customers and visitors.[9]

GPRS is an extension of conventional Global Systems for Mobile communications (GSM) mobile telephony. Slightly faster than GSM (downloading at between 28 and 64 Kbps), GPRS networks offer 'always-on', higher capacity, internet-based services, allowing users to browse the internet, access email on the move, receive multimedia messages and use location-based services. GPRS technologies are predominantly accessed by users of suitably equipped mobile telephones and PDAs (plus the Blackberry mobile email device), but many telecoms operators are offering end-users dual GPRS/WLAN cards for their laptops, which can then be used to roam between the two access technologies. Echoing the coverage available to mobile telephones, GPRS can be used almost anywhere, but users may prefer to switch to the much faster WLAN system once they find they are in a hotspot.

The 3G – 'Third Generation' – communications gives users higher speeds of access than GPRS (itself sometimes called 2.5G), but wide coverage is unlikely to be achieved in the near future. By the end of 2004, only Vodafone and T-Mobile had launched commercial 3G services in the United Kingdom (with Orange and O2 set to follow suit), but coverage outside most major UK cities was expected to be at best patchy until at least the end of 2005. The range of devices using 3G technologies is similar to that for GPRS – that is, 3G-compatible mobile telephones, PDAs and laptops – but the speed of connection, typically around 380 Kbps, may make it attractive to users seeking a high bandwidth wireless alternative to ADSL/SDSL or cable modems – assuming, of course, that it is cost-effective to do so.

BWI – these technologies include WiMAx (Worldwide Interoperability for Microwave Access) and MobileFi, both emerging technologies related to WiFi.[10] WiMax, for example, could potentially enable fast – up to 144 Mbps according to early estimates – wireless connection over an entire city from a single transmitter. WiMAx-based home access points may be available from ISPs and telecoms businesses by 2006, delivering faster 'last mile' solutions than are currently available through ADSL/SDLS or cable; at the same time, WiMAx technology could be incorporated in laptops and PDAs, effectively killing off WiFi and turning urban areas and cities into huge hotspots allowing portable outdoor broadband wireless access (and even free internet telephony instead of paying for mobile calls). In due course, perhaps before 2010, the range of BWI services supported by a single transmitter could extend over a radius of 50 km.

In the construction context, WLANs (and perhaps, eventually, WiMAx) could simplify the installation and management of an IT infrastructure in the temporary office accommodation on site. There would be no expensive cables to lay; adding new users or changing office layouts would be simple; occasional visitors could 'hotdesk' and have easy access to hardware, such as WLAN-enabled printers, scanners and other peripherals. Thinking internationally, long-range wireless solutions would also be valuable in the developing world where terrestrial telephone systems can be rudimentary and unreliable.[11]

In theory, if the hotspot extended out into the construction project site itself, team members could then use the WLAN to access, view and, if necessary, update information held on their construction collaboration system – one could imagine team members perhaps using a hand-held PDA to, say, view a method statement, update a snagging list or even take a digital photograph. For example, WLAN products have already been used by Stent to replace paper-based systems used to monitor piling works at London's King's Cross and Wembley Stadium sites.[12]

Some construction collaboration technology vendors and end-users have already tested mobile devices such as PDAs and hand-held computers on site, but, to date, such trials have tended to focus on asynchronous use – that is, data is not exchanged until the device is placed back on its cradle in the office (however, the author is aware of at least one end-user working for a UK contractor who has been testing out a high-specification PDA equipped with WLAN and GPRS connectivity and using it to access BIW's collaboration system on-site as well as while mobile). Despite a growing market for 'rugged' laptops and PDAs, such devices are unlikely to replace the use of paper on site for briefing site operatives in the near future, but as the technology vendors begin to embrace real-time collaboration more fully, and mobile connectivity becomes more commonplace, one can foresee a time when site-based team members might routinely view project information, make notes, or even hold 'net meetings' with office-based colleagues without first having to return to their site cabins.

10.6   New regulatory pressures

Much of Chapter 7 focused on how existing legal structures needed to be modified to accommodate the introduction of electronic forms of communication within construction teams. However, new regulatory regimes may also have an impact, particularly on large, international corporations.

The impact of the UK's Freedom of Information Act 2000 has already been mentioned, but, following corporate scandals such as Enron, some organisations – particularly if they have multinational operations and deal with financially sensitive information – will find themselves having to comply with the public accounting requirements of the USA's Sarbanes-Oxley Act 2002, the UK's Financial Services and Markets Act 2000 and Companies (Audit, Investigation and Community Enterprise) Act 2004, and similar financially oriented legislation enacted in Europe (e.g. Basel II) and other locations. Such legislation requires corporations to keep a record of every communication entering or leaving the organisation, including paper-based communications, email, instant messages and telephone calls, placing great importance on technologies that allow document capture and retention, records and content management and workflow support, plus storage of the masses of data involved.

Of course, these are not likely to have much, if any, impact on most AEC businesses in the United Kingdom, but they will affect many client organisations and their banking and investment partners. And should regulators such as the American Securities and Exchange Commission or the UK's Financial Services Authority initiate investigations it is possible that communications with suppliers may come under scrutiny. Such regulatory pressures may, therefore, actually encourage customers and suppliers to invest in systems that enable them to capture and store communications. At a corporate level, this may mean heavy investment in email management systems, etc. but there may also be room in the corporate IT armoury for construction collaboration technologies which reduce the load on corporate email systems and divert project- or programme-specific documents and data to their own repository.

10.7 Changing individual working patterns

In addition to the improved efficiency of communication between organisations comprising an existing project team, Chapter 8 described how supply chain SMEs might form 'virtual companies' to compete more effectively. However, there may also be changes and opportunities at the individual professional level.

Wanting to juggle the home/work balance better, or perhaps looking to 'downshift' to a different lifestyle pattern and take advantage of the mobile working opportunities offered by new technology, some AEC professionals have already opted to work as free-lances or as independent consultants, undertaking a succession of contracts of their own choice instead of working for an employer. Particularly in the consultancy sector, just as small firms might combine with others with complementary skills and/or resources, so experienced individual professionals could combine with other independent practitioners to compete for work and then form part of the multi-disciplinary team appointed to undertake the project. Such teams would have a more direct relationship with the customer and this may help customers procuring a succession of projects achieve greater continuity of people, one of the recommendations of *Equal Partners* (Business Vantage 2004a). Being formed of a group of independent 'e-lances' or 'tech-nomads', the operational overheads of such a multi-disciplinary consortium are also likely to be lower, making their services more cost-effective – an advantage likely to be underlined if the team also uses low-cost collaboration technology to manage and share its data.

The increasing ease-of-use and growing reliability[13] of IT, the afore-mentioned growing use of ASP-delivered solutions, and the emergence in many organisations of mobile, home-based workers may also accelerate changes within organisations. As a minimum, one might expect the higher expectations created by ASPs to motivate organisations to demand more of their IT departments and vendors of traditional IT tools, but the change could be even more profound. Futurologist Peter Cochrane, for example, argues that an increasingly transient but IT-capable workforce will cause organisations to reflect on whether they actually need to maintain their own IT systems and attendant IT specialists:

> the next five years will see computers make the same transition automobiles have over the past 25 years – we will not be treating them with any particular reverence. They will become just another commodity, a powerful tool, a convenience, and of course an example of outstanding technological progress that will be taken for granted like all technologies before IT.
>
> What follows this transition? Well, results from a recent survey show that about 80 per cent of office workers across the EU spend at least one day out of the office and some 30 per cent have no office at all. These are the mobile, home-based workers, who most likely don't have a full-time contract with any company. These 'tech-nomads' are on the rise and are [complemented] by growing amounts of out-sourcing across many sectors. Most importantly, they are largely self-sufficient for tech support.
>
> (Cochrane 2004)

However, the gains may not be all positive, unless they are carefully managed and controlled. Recalling old television advertisements for Martini vermouth (any time, any place, anywhere), many IT vendors make a virtue of their solutions becoming increasingly available through the extension of broadband, WiFi, etc. But there are some who argue that the 'always-on' culture can lead to 'connectivity creep' with end-users working longer hours and being on-call 24/7 – at one time, workers could relax offline when travelling on the train or aeroplane, but with the advent of WiFi 'hotspots' on board even those havens are now being eroded. As a result, home and mobile workers, for example, could end up putting in more hours than their office-based counterparts – albeit often willingly – but with little attention paid to health and safety standards.

10.8 The extranet evolution

As suggested at the beginning of this chapter, the gradual adoption of construction collaboration technologies by the UK client organisations and their project teams has been more evolution than revolution. If the ongoing and imminent developments outlined in this chapter are anything to go by, this evolutionary process is set to continue.

To sum up:

- The shift towards more collaborative approaches to the delivery of construction projects and programmes will continue, with prolific clients, particularly those in the public sector and/or those concerned with managing whole life costs, taking a leading role and driving change among their suppliers (though many others will change much more slowly, if at all).
- At the same time, the take-up of collaboration technologies will continue to grow, though there is likely to be a gradual concentration of projects in the hands of a small handful of vendors with the skills, experience and resources to manage large volumes of work. Others may establish niche operations serving loyal customers or, instead of trying to supply all services, may choose to focus on particular aspects of the collaboration challenge (e.g. hosting, software development, training and implementation, system integration, etc.).
- The ASP model shows no signs, yet, of disappearing. Instead it appears to be gathering strength and putting traditional software delivery models under a more critical spotlight.
- Technologically, the software will continue to advance, becoming better integrated with customers' and other end-users' back office systems, incorporating more real-time collaborative functionality, moving beyond sharing of 2D information, and even allowing exchange of information between IT systems and construction components themselves.
- New telecommunications technologies will mean that we no longer need to be physically connected to IT systems to exchange and interact with information. Indeed – and echoing the increasing importance of the ASP model – we may no longer need to carry devices with large and often expensive software applications already loaded, but may increasingly use web-based applications instead.
- Increasingly onerous regulatory requirements will underline the importance of creating and maintaining a full audit trail relating to the exchange of information.
- Last, but perhaps most importantly of all, individual working patterns will continue to evolve. Several factors may come into play – for instance: changing

corporate attitudes to collaboration; better educational and institutional promotion of collaborative working and its supporting technologies; increased awareness, availability and take-up of web-based systems; continuing employee mobility; etc.

It is possible that almost all of the above trends will converge, with take-up of collaboration systems encouraged by a combination of stick and carrot. On the one hand, concerns about corporate compliance will force many organisations to adopt a more stringent approach to capturing and recording all communications sent and received by their employees, while the benefits to be gained from adopting more collaborative approaches will also encourage the adoption of supporting technologies by some project team members. Simultaneously, increased support for mobile working and continued growth of the ASP model will encourage take-up of web-based technologies, which can then be used to interact in real time with an increasingly rich range of data drawn from a more integrated range of sources.

This vision may initially seem a bit far-fetched. But it is worth recalling how much progress the AEC industry can achieve in ten years or so. When the Latham Report was published in 1994, the worldwide web was in its infancy, email was still relatively unknown, e-business was a distant possibility, and the first rudimentary web-based systems for construction collaboration had yet to appear, even in the United States. By the mid 2000s, however, the majority of businesses had their own websites, email had become almost ubiquitous for every employee, e-business had endured a dot.com boom, a slump and at least a partial recovery, and construction collaboration technologies were no longer novel but a normal part of project delivery for a growing number of AEC industry clients, contractors, consultants and other members of the supply chain, in the United States, United Kingdom, Australia and many countries in mainland Europe.

Popular acceptance of technology in general has leapt forward too. We are now more willing to manage our bank accounts online, to book holidays, to buy books, music, groceries and other products and services, and to sell our unwanted items over the internet. New telecommunications capabilities have extended our horizons too. Mobile telephones are no longer solely used for voice communications, but have evolved into more sophisticated devices that offer personal organisers, text messaging, web-browsing, email, cameras, video and music playback and games. Telecommunications providers no longer just supply telephone services, but offer the triple whammy: telephones, television and broadband internet access, all down one cable. And the computer in the living room might at different times be used additionally as a television, digital radio, music player, video recorder, digital editing suite, games console and – with VoIP – even a telephone.

The emergence of wireless technologies could, however, mark the end of the proliferation of separate devices with built-in processors. We may find ourselves using one central unit to connect to the internet, combining the roles of firewall, router, switch, wireless access point and computer, and capable of managing data, voice and other audio, and images and videos. Depending on our different needs, we could then use a variety of simple interface devices that send and receive data from this central unit for telephony, audio and video entertainment, office communications, etc.

This relative profusion of IT functionality in the office, home and pocket or briefcase is a strong indicator of just how reliable, easy-to-use and affordable IT has become. The fact that construction collaboration technologies have yet to become so widely adopted – particularly on smaller projects – is probably an indication that they have yet to become as reliable, easy-to-use and cost-competitive. However, this may simply reflect the relative immaturity of the sector.

In terms of reliability, construction collaboration technologies are an advance on systems that were largely paper-based and contained much scope for human error and logistical breakdowns. And, compared to many internally managed systems, remotely hosted ASP solutions tend to be more available, more transparent and less prone to service interruption. But this reliability is not universal. In some instances, the infrastructure can be a weak link, with, say, poor internet connectivity hampering efficiency; in other cases, the weak links might prove to be human, with team members resistant to what they see as an over-transparent technology or simply prone to making mistakes.

This brings us neatly to ease-of-use. Understandably, most of the existing vendors have focused their market-building efforts on winning over the sophisticated clients and teams involved in large, complex projects and programmes of work. Satisfying such customers that the vendor's systems will meet the team's needs has led to the development of rich levels of detailed functionality, but these can often appear too complicated when considered by more down-to-earth clients and teams involved in much smaller and/or simpler schemes or by those who still want to host their own systems. Some vendors are addressing ease-of-use issues by designing simpler interfaces and reconfiguring their applications so that they can support simple processes, but there is still scope for improvement. Will existing products be re-engineered to appeal to the wider AEC market at the lower end of the project scale? Will existing vendors perhaps develop simplified versions of their systems that they will continue to host? Or will new vendors target the AEC market for self-hosted, simple collaboration? Perhaps only time will tell.

If construction collaboration technologies are to become adopted more widely, they will also have to satisfy its customers and end-users as to their affordability. As described in Chapter 9, many benefits are claimed for the technologies, but the costs can sometimes appear prohibitive. It is perhaps in this area, more than any other, that the vendors and other proponents of the technologies can do most to tackle resistance. If they can build up a convincing business case for the technology which explicitly and objectively quantifies the net savings and other benefits to be achieved, then prospective customers and end-users should need little persuasion on cost grounds to make the change.

10.9   Chapter summary

This chapter – the final chapter in this book – was a look to the future, speculating about trends in the continued cultural and corporate evolution of the AEC industry in the United Kingdom, the continued development of its IT tools and telecommunications, and ongoing and future changes in the attitudes and working practices of its people. It expanded on many points raised in earlier chapters, identified several issues that may (or may not) influence the future development of the UK market for construction collaboration technologies, and suggested how these trends might converge. It is, perhaps, an optimistic account. It anticipates that clients and the industry will want to continue to change, that organisations and the people within them will become more committed to collaborative approaches to projects, and that the technologies and their vendors continue to evolve to support such approaches. It also imagines that there will be a trickle-down effect that will see collaborative working and its supporting technologies increasingly adopted at all levels of the UK construction industry, not just by progressive organisations at the larger end of the scale.

The author may, of course, be proved wrong on many, if not all, points. But, after some 18 years in the UK construction industry, he remains hopeful that there will be progress in at least four key areas:

- less reliance on slow, labour-intensive paper-based processes to share project information;
- greater integration between design and construction people, processes and technology;
- more transparent, long-term commercial relationships;
- an improved industry reputation for efficiency, reliability, safety and innovation.

One thing must be clear: standing still is not an option. To use a phrase attributed to 'Anon' but often repeated by Sir Michael Latham: 'If you always do what you always did, you'll always get what you always got.'

Glossary

ADSL During the early years of the twenty-first century, asymmetric digital subscriber line (ADSL) has become increasingly popular as a means of bringing high bandwidth information to homes and small businesses over ordinary copper telephone lines, without requiring a separate voice line. It is called 'asymmetric' because most of its two-way bandwidth is devoted to the downstream direction, sending data to the user – usually at up to 512 Kbps – as opposed to sending data from the user – usually at up to 256 Kbps. Some providers also provide higher capacity ADSL services, for example: 1 Mbps.

AEC Architecture, engineering and construction.

ASP An application service provider (ASP) is a company that offers individuals or enterprises access via the internet to applications, data and related services that would otherwise have to be located in their own personal or enterprise computers. The software is essentially rented by the customer for as long as it needs it.

Audit trails An audit trail records every transaction related to a particular document, drawing or process. Detailing who did what and when, an audit trail means users can no longer claim non-receipt of information, etc.

B2B Business-to-business (B2B) is used to describe the exchange of products, services or information between businesses via the internet.

Bandwidth Bandwidth is the ability of a telecommunication component to deal with an amount of electronic transmission of any digital communication between two points. The higher the bandwidth you have, the faster data can be transferred.

Blogging Short for 'web log', a blog is a personal journal that is frequently updated and intended for general public consumption. Topics sometimes include brief philosophical musings, commentary on internet and other social issues, and links to other sites the author favours. Although initially mainly confined to internet-savvy enthusiasts, some businesses have also starting to consider its potential, perhaps as part of an organisation's knowledge management strategy to tap into the specialist interests of its employees.

Broadband A high-bandwidth type of internet connection. The term broadband has been abused to refer to quite slow connections, but most commentators regard broadband as referring to rates of 256 Kbps upload/512 Kbps download or greater.

Browser A browser is an application program that allows users to access, view and interact with information on the WWW. The most widely used is

Microsoft's Internet Explorer; others include Netscape Navigator, Mozilla Firefox and Opera.

CAD   Computer-aided design or computer-aided drafting (CAD); also known as computer-aided drafting and design (CADD). CAD software is used by architects, engineers, artists and others to create precision drawings or technical illustrations. CAD software can be used to create two-dimensional (2D) drawings or three-dimensional (3D) models.

Client   On a client–server architecture, a client is the application on a user's own computer that allows the user to exchange data and interact with the server. A web browser is an example of client software.

DDoS   A distributed denial of service (DDoS) attack can be instigated by a hacker or by criminals intent on extortion. Using hundreds, even thousands, of computers around that world that have been infected with malicious programs, the instigators can bombard a website's servers with requests, thereby using up most if not all of the available bandwidth. The website's performance can slow down or even grind to a halt.

Domain name   A domain name is the part of an email or website address that details the name of an organisation or company, what type of organisation it is and in some cases the country in which the company is located.

EDMS   Electronic document management system.

Extranet   Essentially, an extranet is a private network that uses the internet and the public telecommunication system to securely share a business's information or operations with authorised suppliers, vendors, partners, customers, or other businesses. An extranet can be viewed as an extension of a company's intranet (i.e. a private network contained within an enterprise – see p. 190) for users outside the company. By this definition, so-called 'project extranets' are not strictly extranets as they only provide access to a subset of the enterprise's knowledge- or information-base, usually specific to the project(s) concerned.

Firewall   A firewall is a piece of hardware or software used for security purposes to filter all data traffic between a computer (whether an individual PC or a corporate server) or LAN/WAN and the internet. It can, for example, help prevent users receiving viruses and other unwanted email file attachments, and prevent unauthorised access to an individual's or organisation's computer systems.

Groupware   Groupware refers to programs that help people work together collectively while located remotely from each other. Services can include the sharing of calendars, collective writing, email handling, shared database access, and electronic meetings with each person able to see and display information to others. Product examples include Lotus Notes and Microsoft Exchange, both of which facilitate calendar sharing, email handling, and the replication of files across a distributed system so that all users can view the same information.

Hardware   All the tangible, physical components of a computer system, from keyboards, mouse and monitors, to internal components such as hard-drives and chips to peripherals such as printers.

HTML   Hypertext mark-up language (HTML) is the simple programming language used to create many web pages.

ICT   In some sectors, information and communication technology (ICT) is increasingly preferred to 'information technology' or IT, as it reflects the increasing importance of the internet and its supporting telecommunications infrastructure to the delivery of information. IT remains the more commonly used term within the UK construction industry, and is the term used by this book.

Internet   The internet is a global system of interconnected computer networks. Initially conceived as part of a US Government research project and known during the early 1970s as the ARPAnet, it evolved into a vast public network.

Interoperability   This term describes the ability of systems to interact with each other. For example, interoperable systems allow information from one application (e.g. CAD) to be shared with another (e.g. a database).

Intranet   An intranet is a private network contained within an enterprise. It uses IP and its main purpose is normally to share company information and computing resources among employees.

IP   Internet Protocol (IP) is the protocol employed to send data from one computer to another via the internet. Each computer has at least one IP address that uniquely identifies it from all other computers. When you send or receive data (e.g. an email or a web page), the message is divided into little 'packets'. These may arrive at their destination in a different order to which they were sent, but the Transmission Control Protocol (TCP) puts them back in the right order.

ISDN   Integrated Services Digital Network (ISDN) is a set of standards for digital transmission. ISDN lines were considered more efficient at data transmission than ordinary telephone lines, but the advent of ADSL/SDSL and cable modem services means ISDN is now less popular than it once was. Basic ISDN services operate at up to 128 Kbps, while primary services can operate up to 15 times faster: 2 Mbps.

ISP   An internet service provider (ISP) is a company that provides individuals and other companies access to the internet and other related services such as website building and hosting. Options include: a dedicated server leased from the ISP; sharing a server with other (sometimes competing) applications and/or websites; and co-location – the provider places its own server, set up with all the software, in the ISP's facility.

IT   See ICT earlier.

LAN   A local-area network (LAN) is a network connecting computers and other devices in one office or building.

Metadata   In an IT context, the prefix 'meta' means 'an underlying definition or description'. Metadata is therefore a definition or description of data. In the context of construction collaboration technologies, it might be helpful to think of it as an 'intelligent wrapper' around a document, drawing or other item that contains details about who sent the item, when, and to whom, its status or priority, drawing number, title, revision code, etc.

Modem   A shortened form of 'modulator/demodulator': the device used as an interface between a computer and the telephone system. Thus, 'inward', the modem would receive and convert audible sounds from a telephone line into the digital signals used by a computer, and, 'outward', would convert the computer's digital signals into audible sounds for telephone transmission.

Peer-to-peer   Peer-to-peer (P2P) is a kind of transient internet network that allows groups of computer users with the same networking program to connect to and directly access files from each other's hard drives. While P2P became particularly well known for downloading music and other files (e.g. through Napster and Gnutella), it also has great potential as a business tool, offering a way for employees or businesses to share files without the expense involved in maintaining a centralised server.

Plug-in   Plug-in applications are programs that can easily be installed and used as part of a standard web browser. They are recognised automatically by the browser and their functions are integrated into the main HTML file that is being presented. For example, Adobe's Acrobat Reader allows users to view PDF documents just as they look in the print medium. In the context of construction collaboration technologies, most systems' drawing viewers are plug-ins.

Protocol   A set of rules for sending data, usually across a network or the internet.

Proxy server   A proxy server acts for computers on a computer network as a gateway on the Internet. It receives requests for web pages and goes online to get them, then sends the information back to the computer that made the request. Due to all the traffic flowing through the proxy server, all the traffic can be monitored and controlled, and user access can be filtered.

Redlining   Marking up drawings electronically on screen with comments (e.g. to suggest changes, ask questions, etc.).

Router   A router is a piece of hardware that sends and receives traffic between networks. They can be used to plug LANs together with leased lines.

Scalability   The ability to increase a system's capacity or number of users without huge upheaval or expense.

SDSL   Symmetric digital subscriber line (SDSL) is an alternative to ADSL (see earlier). It also offers high bandwidth internet connection to homes and small businesses over ordinary copper telephone lines, without requiring a separate voice line, but is 'symmetric' insofar as both upstream and downstream data transfer are equally fast – say 512 Kbps. Again, some providers also provide higher capacity SDSL services, for example, 1 Mbps, 2 Mbps, etc.

Server   A server is a central computer that awaits and fulfils requests from other computers (or 'clients') on a network. For example, someone using an EDMS would be requesting information using a client program on their laptop or desktop machine, and the request would be fulfilled by an EDMS server within the same LAN/WAN. A web server essentially performs the same function; requests are made by a web browser client, but fulfilment is typically accomplished from *outside* a user's LAN or WAN.

SLA   Service level agreements (SLAs) specify in detail how an ASPs service might be delivered.

Software   Software is the term used to describe intangible elements of a computer system, for example, data, applications, programs and instructions.

Spam   Spam is unsolicited email on the internet. From the sender's point of view, it is a form of bulk mail; to the receiver, it usually seems like junk email. Spammers typically send a piece of email to a distribution list in the millions, expecting that only a tiny number of readers will respond to their offer. Spam has become a major problem for all internet users. The term is said to derive from a famous *Monty Python* comedy sketch (Well, we have Spam, tomato & Spam, egg & Spam, Egg, bacon & Spam…) that was current when spam first began arriving on the internet. SPAM is a trademarked Hormel tinned meat product.

TCP   See IP.

Virus   In computers, a virus is a program that replicates by being copied or initiating its copying to another program, computer or document. Viruses can be transmitted unwittingly as attachments to emails or in downloaded files, or be present on disks. Some viruses activate as soon as their code is executed, others lie dormant until circumstances cause their code to be executed. Many viruses can be quite harmful, erasing data or requiring hard disks to be reformatted. A virus that replicates itself by resending itself as an email attachment is known as a worm.

VoIP   Voice over IP (VoIP) is a term used to describe facilities for managing the delivery of voice information using the IP (see IP). In general, this means sending voice information in digital form in discrete packets rather than in the traditional circuit-committed protocols of the public switched telephone network. A major advantage of VoIP and internet telephony is that it avoids the tolls charged by ordinary telephone service.

VPN   A virtual private network (VPN) is a secure private network using a public network (e.g. the internet) as carrier.

WAN   A wide-area network (WAN) is a network connecting computers and other devices in geographically dispersed offices or buildings.

Webcam   A webcam is a video camera, usually attached directly to a computer, whose current or latest image can be viewed on a website. Live webcams typically provide a rapid succession of new images or, in some cases, streaming video – assuming adequate bandwidth.

Wiki   A wiki (the term comes from the Hawaiian word for fast: 'wikiwiki') is a server program that allows users to collaborate in forming the content of a website. With a wiki, any user can edit the site content, including other users' contributions, via a standard web browser. Basically, a wiki website operates on principles of collaborative trust and constant peer review (one of the most well-known public projects is the Wikipedia, an online encyclopaedia, but wikis have also been used within organisations as collaborative knowledge management tools). A 'Twiki' is similar but allows users to track the changes made by other users; revision control archives all previous content so nothing is lost.

XML   Extensible markup language (XML) is a flexible way to create common information formats and share both the format and the data on the WWW, intranets, extranets, etc. It can be used by any individual or group of individuals or companies that wants to share information in a consistent way.

Notes

1 Defining collaboration

1 At one time, the word internet was often used with an initial capital letter: Internet. However, as the term has become more widely used, this convention has begun to lapse. Several UK newspapers and magazines including the *Daily Telegraph*, the *Independent*, the *Guardian* and the *New Scientist* all now use the lower case form, and this book employs the same convention.

2 Reflecting the increasing importance of the internet and its supporting telecommunications infrastructure to the delivery of information, the term 'information technology' (IT) is often expanded to 'information and communication technology' (ICT). Most construction organisations still tend to use the IT abbreviation, however, and this book will follow that convention.

3 A poll in US magazine *Engineering News Record* (21 June 2004) put the internet at the top of its top 9 electronic technologies that changed construction, ahead of CAD, lasers, analysis software, PCs and the fax.

4 Some UK vendors have regional businesses in mainland Europe, most notably BuildOnline. Others have tended to venture into Europe only when required by their customers.

5 Many vendors have jumped on the collaboration bandwagon. An advertisement in a leading IT magazine once invited businesses to enable collaboration with a powerful electronic communication device. The vendor said the product would bring multiple parties together in real time, perform outstandingly, and offer a 'low cost, easy deployment' method to cross company boundaries. It was a speakerphone.

6 For example, some of the leading global names in IT offer 'enterprise portal' or 'intranet' applications (e.g. Microsoft, IBM, Oracle, SAP, Computer Associates, Sun, PeopleSoft), along with a host of more focused portal specialists (e.g. Plumtree, Autonomy, Documentum, Vignette, OpenText). Then there are vendors of generic workspace or project team applications (e.g. Groove Networks, ProjectPlace), web and video conferencing and online meeting applications (e.g. WebEx, Microsoft's NetMeeting), real-time instant messaging (IM) (e.g. Yahoo, AOL) or 'chat' applications, presence awareness, e-learning, etc.

7 In the absence of any similar UK-based publication, Joel Orr's influential American website (www.extranetnews.com) and popular online newsletter *Extranet News* probably helped perpetuate the term.

8 For example, Michael Thompson of Team Focus told a Salford Centre for Research and Innovation seminar on 13 July 2004: 'only about 30 per cent of the construction industry is practising partnering or collaborative working in any shape or form. Instances of full integrated project teams are probably far less than this – probably less than 10 per cent'. A *Contract Journal* poll ending on 1 December 2003 asked 'Are contractors and clients adopting the correct methods of partnering?' to which 79 per cent answered no. A later poll, ending 25 May 2004, asked 'Is collaborative working actually driving through change in construction projects?' to which 60 per cent answered no. On the other hand, a pan-industry telephone survey undertaken by Collaborating for the Built Environment (BE) in late 2004, found 70 per cent of construction businesses rated collaborative working

as important (7 or above on a scale of 1–10) to their future success, and just over half (51 per cent) rated their last completed project as integrated.

2 The convergence of culture and technology

1 In July 2004, it was announced that BE was in convergence discussions with Constructing Excellence (itself the result of merging Movement for Innovation (M4I) and the Construction Best Practice Programme (BPP) in 2003) and the Construction Industry Research and Information Association (CIRIA). By late 2004, CIRIA dropped out of the discussions and the merger of BE and CE was completed in early 2005.

2 This followed a similar project in 2002 (Business Vantage 2002) that examined customer and supplier alignment in private sector construction. In 2004, Business Vantage also revisited the private sector contributors to the original research and produced a further report on progress since 2002 (Business Vantage 2004b). This noted that 'the use of web-based technology in design, collaboration, procurement and knowledge management offers potential productivity and efficiency improvements. Customers in 2004 believe that more is being invested in IT although the message is clear that many could do more' (p. 31).

3 A survey in US magazine *Engineering News Record* (15 May 2000) suggested that construction ranked 87th among industries globally in adopting technology.

4 Recalling the 1980s, *Building* magazine (18 June 2004) quoted Ian Hamilton of the Construction Industry Computing Association: 'An IBM PC could do useful things and it would only cost a couple of thousand pounds – so medium-sized organisations could afford them.' Today, a firm spending that amount of money could afford a PC that is about 120 times faster. It also quoted the CICA's Erik Winterkorn: 'In the 1980s, at best you had one CAD workstation for every three users.... Twenty years ago, £250,000 would have bought a couple of workstations, a processor and a plotter; now a high-spec PC can do the same job' (p. 42).

5 Construction industry personnel were among the most enthusiastic early adopters of mobile telephones and of faxes.

6 A Cardiff University survey of architectural practices undertaken by Mike Fedeski and Bhzad Sidawi in 2000 found only 1.7 per cent of firms were connected to the internet in 1994.

7 The Building Information Warehouse (BIW) project marked the beginning of BIW Technologies as a company (the portal is still managed by BIW, at www.biw.co.uk). However, subsequent research projects shifted the company's focus more towards solutions allowing construction professionals to share project information online.

8 By 2004, 80 per cent of construction businesses had a website (DTI 2004).

9 Source: ResearchWorldwide.com press release issued 14 December 2004.

10 A 'killer app' is an application so useful that it surpasses all others and even drives people to buy a computer simply to run that application.

11 An email study headed by Tim Moors at the University of New South Wales (UNSW), Australia (reported by Brian Livingston on http://itmanagement.earthweb.com/columns/executive_tech/article.php/3404381 on 7 September 2004), found that 1.57 per cent of emails did not reach their recipients due to 'the flaky nature of Internet servers, clients and routers'.

12 One research firm estimated the average employee received around 7,500 spam messages a year, while handling this deluge took 3.1 per cent of an employee's time, equivalent to £775 for every £25,000 of salary. Research from the Royal Bank of Scotland Corporate (reported June 2004), which questioned 1,000 small businesses, found one in 10 small and medium-sized businesses believe they spend £10,000 a year dealing with spam.

13 Becerik (2004b) gives a good account of the development of the US market.

14 An ASP is a company that offers individuals or enterprises access via the internet to applications and related services that would otherwise have to be located in their own personal or enterprise computers. In the UK AEC industry, some vendors also provide their software to run on a customer's own computers; it is debatable whether they are true ASPs, but for the purposes of this chapter, ASP is taken to include them. (It is also worth noting that some

applications for the AEC market are not 'pure ASP' insofar as they need users to install or download a client program before full functionality is possible.)

15 BuildOnline launched initially in Ireland as an e-marketplace before relocating to the UK and focusing on collaboration.

16 Notably Emap's ConstructionPlus Project Extranets site at www.cnplus.co.uk/proj_collaboration/?ChannelID=35 (last accessed 27 May 2004).

17 These client organisations are in the minority insofar as they procure a steady stream of construction projects. The majority of the UK non-residential market is dominated by construction industry clients who are 'one-off' or 'occasional' buyers. The prolific clients were in a strong position to drive early adoption and to create exemplars and standards for their more 'occasional' counterparts. Where they also managed their own capital assets, they could also drive use of the collaboration technology for its later value in supporting facilities management (FM) activities.

18 Asda later decided to focus on use of the Sarcophagus system.

19 The value of some surveys has been limited by poorly constructed or poorly worded questions, or by unrepresentative samples. In the Construction Confederation (2001) survey, for example, 'using the internet for project collaboration' could also include use of email as a tool for collaboration, inflating the numbers beyond those employing a construction collaboration technology. Indeed, the Confederation warned: 'some of the figures are on the high side and may have been distorted by how the question was phrased or by the sample that chose to respond to the survey'.

20 Such figures suggest the UK has lagged behind the US market in its adoption of collaboration technology. A 2001 survey by the US Construction Financial Management Association (CFMA) showed 40 per cent of general contractors had tried construction collaboration software; a poll by the Associated General Contractors of America in 2002 said 52 per cent of respondents used online collaboration. Another US survey, by Ahmed *et al.* (2002), suggested that 59 per cent of firms (drawn from contractors, construction and project managers, and architects and engineers) were using project management or project collaboration websites to manage their projects. A 2004 CFMA survey (summarised by Sawyer 2004), showed that 64.5 per cent of contractors earning over US$250 million per annum were using collaboration software.

21 Seizing on the Barbour (2003) findings, UK vendor Cadweb issued a press release trumpeting: 'Of those who use extranets just short of one in four use Cadweb. Less than one in ten use the next most common system. This survey indicates that Cadweb's market share is more than $2\frac{1}{2}$ times greater than its next biggest rival, clearly illustrating the gulf in market share that exists between Cadweb and its competition.' But the Cadweb claims were based on less than 20 survey responses, some clients may have confused CAD software with Cadweb's application, and its claim to be 'undisputed market leader' (quickly disputed by other UK providers) was undermined by its own figures and by other market research (e.g. Compagnia 2003).

22 Reported in Construction Industry Computing Association (CICA) Bulletin 91, Winter 2004, p. 7.

3 The construction collaboration providers

1 'The List' at Joel Orr's www.extranetnews.com website, for example, included companies targeting manufacturing, mechanical engineering, healthcare, legal and packaging sectors, while the products included many non-extranet solutions, e.g. portal systems, bandwidth management tools, product development support, online forums and discussion groups, blogging and IM software, time-tracking, file translation and file viewing utilities, news services, calendar and task-list sharing tools, wiki's, etc.

2 This chapter is not intended to give a detailed and comprehensive overview of all the providers. It is acknowledged that there may be other providers who are not mentioned or not described in more detail. Readers will also be aware that the IT world is volatile; some of the businesses mentioned may have ceased trading, undergone a change of ownership, or ceased delivering the applications mentioned.

3 In the United States, Becerik (2004a) found a common problem among current solutions was that 'most of them aren't built by people with construction knowledge and experience... Some of the modules and processes...were...poorly designed and insufficient. The tools should be developed in conjunction with construction experts, so that their understanding of the way work needs to be done in AEC industry can be built into the tools' (p. 5).

4 Used initially as a knowledge management application for internal collaboration, global engineering and construction firm Montgomery Watson Harza (MWH) has also used Lotus Team Workplace (QuickPlace) to manage UK construction projects.

5 Hummingbird was so named to reflect Bovis's bird logo, and should not be confused with the Canadian software company of the same name, some of whose products are also described as collaboration tools.

6 To support Bovis's business processes, the server-based Hummingbird system has other modules that complement the Image Management function, including financial control and tendering management. Within Bovis Lend Lease's Asia Pacific region, the company also uses ProjectWeb, a Lotus Domino browser-based system (developed with Lend Lease by Australian IT firm Didata) that uses Adobe Acrobat PDF files for drawing images.

7 'Vapour ware' was rife during the dot.com boom, and not surprisingly AEC buyers were urged to be cautious in choosing who to do business with. For example, Kinns (2000), for the Institution of Civil Engineers, suggested various questions that potential buyers needed to ask, his first ones indicating the scepticism that existed: 'Do they have a working product? Have they got financial backing to ensure they will be around at the end of the market shake down?'

8 Several other equally notable US-based vendors remained focused on their home market and – at the time of writing – remain relatively unknown in the United Kingdom (e.g. Constructware, e-Builder, Tririga).

9 Just as some American providers focused on the home market, several European businesses did not look beyond their own borders, for example, ByggeWeb in Denmark (Howard 2002), Raksanet in Finland, and Conject in Germany.

10 Founded in Israel in 1998, I-Scraper folded in early 2001, its United Kingdom and German operations being taken over by BIW and BuildOnline respectively.

11 As already mentioned, AEC/communications GmbH's product is resold in the United Kingdom by Asite.

12 As mentioned, AEC business Arup, for example, started offering collaboration services, but later off-loaded its applications to Causeway Technologies.

13 When comparing independent vendors with collaboration vendors operating as subsidiaries of larger businesses, potential customers should also look carefully at how the overheads of the latter are managed and accounted for. A business could appear to be profitable partly because its accommodation, IT, human resources, accounting and other support expenses are shared with the parent business.

14 There were also fears, particularly with respect to providers entering the UK market from overseas, that project data might not be held in the United Kingdom but in a central repository in, say, California, with different legal safeguards over data protection, security, etc.

15 The latter is a moot point, as the cost-effectiveness of the solution might not be the same for other users if the backer was getting reduced prices in exchange for permission to use its name in publicity by the preferred provider. Moreover, where a vendor may have offered more than one collaboration solution (e.g. it may have been a re-seller of more than one product), or the collaboration solution was part of a wider package of software and/or hardware and/or services (e.g. collaboration was a feature within a bigger enterprise portal system), it may not be clear whether the benefits arose directly or solely from the particular package now being offered.

16 Of course, if an ASP is hosting software provided by another vendor, the model will differ slightly. Typically, such an ASP might pay an up-front licence fee to the vendor and strike a long-term maintenance deal to obtain upgrades and technical assistance from the vendor. As new customers seek to rent the software, the ASP may also pay a commission fee to the vendor, but will earn the bulk of its income from the services and support it offers for the duration of the customer relationship. Thus, any revenue will not be solely the ASPs. For

example, in fixing its customer charges, Causeway Technologies would have to take into account any payments it makes to OpenText to use its Livelink software.

17 Deferred revenue becomes an important financial metric, indicating how much service the company has left to deliver on its contracts, and future financial performance is less volatile and more predictable. It takes longer for an ASP to become profitable, because, for every deal made, it only gets a small proportion of the revenue at a time, having deferred the bulk of the revenue instead of getting it all up-front; in the meantime, it has also had to make substantial investments in its software and hosting facilities. But once it becomes profitable, it is very profitable, as the ASP has money coming in every month from past deals.

18 There are variations in between these two models. For example, in late 2004, SAP began to offer its enterprise software on a 'pay-as-you-go' model, offering its applications for £325 per user, including installation and maintenance costs. However, SAP admitted it was not technically offering subscription software; instead it was spreading the cost of its software through a special three-to-five year financing arrangement.

19 For example, in 2004, Autodesk announced that from January 2005 it would no longer be supporting AutoCAD 2000i and related products. A survey by a rival CAD vendor (Bentley) revealed that 87 per cent of AEC users objected to being forced by Autodesk to upgrade or retire their software.

20 In addition to the monthly subscription, a customer will usually also expect to pay some implementation costs to cover consultancy, configuration and training. Ongoing support is normally covered by the subscription payment.

21 For example: for hosting, one UK vendor's price list showed it charged £1,000 for installation and configuration, £15 per seat per annum (including 25 MB disk space), with each additional 1 Gb costing £600 per annum.

22 While the US market has seen a number of mergers and liquidations, etc. the UK market has, to date, only seen one major casualty, iScraper. However, Bidcom has in effect been swallowed up by E-box (at least three former Bidcom executives moved to Asite), while e-Hub.com stopped marketing project collaboration services, being focused on e-procurement. As a result, there has been relatively little market consolidation and market leaders have been slower to emerge.

4 Hosting construction collaboration technologies

1 Sometimes also called 'hosted application vendors'.

2 However, research by Nitithamyong and Skibniewski (2004) found that success or failure of a solution did not depend on whether a system was internally hosted or hosted by a service provider. For example: 'Projects using internal servers ... did not show any significantly higher benefit regarding the security of data, which could also mean that security systems provided by most service providers are relatively reliable' (p. 23).

3 The ongoing costs of R&D, support and hosting of an ASP-delivered service are also more transparent. While some room for negotiation will always remain, customers will not be able to argue that, because the vendor has already done the R&D, written the code and pressed the CDs, that the incremental cost of software delivery is minimal.

4 Scott Unger is President and CEO of Constructware and is perhaps ideally placed to comment, having experience of both types of applications within the AEC sector. He developed a client–server based project management application in 1994, and then introduced what was claimed to be the industry's first hosted ASP solution in November 1997.

5 Here, security does not just relate to keeping hackers and viruses at bay, it can also relate to the physical security of an IT facility. There is often a greater risk of someone breaking into a conventional IT department and stealing a server or other equipment than there is of someone doing the same at an ASPs hosting facility.

6 Given the AEC sector's relatively low level of investment in IT, sunk costs should be less of an issue for most construction firms.

7 During the dot.com boom, numerous ASPs announced their intention to seek funds through an initial public offering (IPO), but this trend evaporated when the technology markets crashed. However, during 2004 there were signs that market confidence in ASPs

was changing; in June 2004, US customer relationship management ASP salesforce.com raised $110 million in an IPO, and RightNow had announced its intention to IPO.

8 Reported in *The Observer* newspaper, 5 December 2004, p. 7 of Business section.

5 Features and functionality of construction collaboration technology

1 There can be exceptions. Some client organisations do not want to use the internet or external hosts, usually for security reasons. A Bristol-based architect undertaking work for a major UK bank described to the author how the client's project data, on its in-house developed system, could only be accessed from a particular computer in his office via a dedicated secure communications link.

2 Breetzke and Hawkins (2003a) summarises responses to a survey undertaken for the RICS in 2002.

3 Such pre-qualification questionnaires need to be quite detailed, and should be prepared by someone with a good understanding of the IT sector, collaboration applications, and the types of businesses offering services. A clear brief must also be given. Otherwise, it can be very difficult to compare like with like (see also discussion of reverse auctions in Chapter 8).

4 Clearly, it is the client and each project team company's responsibility to grant logins only to authorised users, and to ensure that these users are both fully aware of the security requirements and will take appropriate precautions to safeguard their log-in details. CITSEC (2004), for instance, suggests a four-point regime: 'DON'T share your password; DO log off when finished; DON'T leave a logged-on machine unattended; REMEMBER that you are personally responsible for anything done using your account.'

5 The distinction between folders-based systems and registers-based systems (i.e. those using a relational database) is often a subject of fierce debate between both users of the different systems and the providers themselves. The key requirement is that, whatever system architecture is employed, end-users must feel confident that they will be able to configure the system to suit their immediate and future project needs and then be able to navigate the structure and find what they are looking for quickly and efficiently.

6 Some collaboration systems offer integration with existing electronic drawing management systems so that an architect, say, would not need to type in the same information twice.

7 A CAD managers survey (Davies 2004) highlighted the need for the AEC industry to reduce the amount of time spent formatting and translating CAD information for distribution (p. 15). Responsibilities were allocated to both software developers/vendors and AEC companies. The former, it said, needed to place more emphasis on simple 2D 'interoperability' rather than using file format changes as a source of continued revenue and competitive advantage, while all AEC companies needed to streamline and rationalise their issuing procedures so that software vendors could better understand the general needs and provide the necessary tools.

8 In compiling project protocol documents, teams may wish to consider what drawing sizes they wish to use, and relate these to the reproduction facilities each team member has at its disposal. Some project teams may standardise on A3 prints for local use on site, for example, using a laser printer rather than a more expensive plotter.

9 While it is often claimed that construction collaboration systems cut paper and printing costs, some project team members may feel that the burden has, in fact, merely been handed down to recipient by the issuer. There will almost certainly be some savings overall (many of those included on an issue list do not always have to print out a full-size copy of the original), but project team members, particularly smaller contractors, consultants or suppliers without their own plotting facilities, may need to make arrangements to either receive paper copies or use a bureau service (e.g. ServicePoint, Hobs).

10 Particularly for professionals used to checking traditional drawings on paper, using an electronic system can seem onerous and difficult. Some recipients may feel that their work requires sight of the whole drawing rather than just the portion viewable within the collaboration system's viewer. Also, in some instances, recipients may want to consult with colleagues in their office before making comments. In both instances, they may end up

printing the drawing for checking purposes and then copy their comments or amendments on to the screen version.

11 In some instances, designers will use vector graphics during creation, but may save the file as a raster image or as a PDF so that others can view it but not edit it. Such conversions can result in minor changes to fonts, colours, etc.

12 Through PDL a computer specifies the arrangement of a printed page to a printer. A PDL defines page elements independently of printer technology, so that a page's appearance should be consistent regardless of the specific printer used. The printer itself (rather than the user's computer) processes much of the graphical information. For example, the printer carries out a command to draw a square directly rather than downloading the actual bits that make up the image of the square from the computer.

13 CAD-to-PDF conversion programs include Jaws PDF Creator and Softcover's vector graphics-based Plot2PDF. Apple Mac OS-X users can generate PDFs by selecting it as a print function, as can users of Bentley MicroStation V8 2004. Note: some PDF creators do not create true vector files, instead they create a bit-map (e.g. JPG) of the drawing and embed it in a PDF (usually with a large file size).

14 For example, Square One's pdf2vector enables PDF files to be imported and edited in AutoCAD and other engineering applications, helping designers to reuse drawings that have been archived and exchanged in PDF format.

15 Multi-threading is the ability of a program to be used by more than one user at a time or to manage multiple requests by the same user without needing multiple copies of the program running in the computer.

16 Autodesk's Volo View Express was a lower specification, 'freeware' version, lacking Volo View's heavyweight red-lining, precise measuring and higher quality plotting capabilities.

17 This paragraph summarises developments over several years. As with many applications, the pace of change is unlikely to slacken in 2005 and beyond.

18 Some also support hundreds of file formats, even though most well-run projects tend to standardise on just a handful of file types.

19 In August 2004, however, BIW announced it was to work with netGuru's Web4 division to add new functionality to its in-house developed viewer.

20 For example, while the vendors using, say, AutoVue may trumpet that their viewer supports over 200 different file formats, most project teams decide to standardise on just a handful of common formats so the wider level of support may be irrelevant.

21 This last point is a vital one, and relates to the discussion in Chapter 3 about choosing a software vendor who will be around to offer support on its collaboration system for a substantial period of time.

6 Connecting to a construction collaboration service

1 There are still some organisations which have either resisted the internet revolution or which have strong operational reasons for not connecting to the internet. Certainly some government departments or agencies have strict guidelines about its staff accessing the web from their internal networks; for a project for the UK Ministry of Defence at Andover North, for example, client members of the project team used stand-alone PCs – outside the MOD's secure networks – to monitor and contribute to team collaboration.

2 The precise requirements of each business (and perhaps each business location) will vary. Readers are recommended to seek expert guidance on the most appropriate hardware and software infrastructure for their purposes.

3 The 11 September 2001 attacks on New York's Twin Towers and on Washington led to all major news websites becoming swamped and unable to return page requests.

4 At one time, UK web surfers would find the internet significantly fast in the morning, but its performance would deteriorate rapidly as the United States came online. Accordingly, some early UK collaboration users did complain, for example, it takes too long to acess the system, especially after lunchtime.

5 One well-known filter package incorporates a database with over 10,000,000 URLs which is scanned when checking content.

6 One leading ISP, AOL, fell foul of the UK Advertising Standards Authority (ASA) in June 2004 about advertising claims made for its 1 Mbps broadband service. A complainant argued that, because the service was shared between lots of different users, it was unlikely to reach 1 Mbps for downloads. AOL admitting there were various hurdles – including the make and model of PC, the distance between the user's home and the local BT exchange point, the time of day and the number of users online – but said these were beyond its control, yet if everything was going their way at any given point, the speed was achievable. The ASA adjudicated that AOL's advertisements had exaggerated the likelihood that such speeds could be achieved; AOL started to use the phrasing 'up to 1 Mb' in its promotional material. It pays to be wary of ISP claims; two months later (August 2004), ISP Wanadoo was also criticised by the ASA for advertising its 512 Kbps service as 'full speed broadband' when other, faster services were also available.

7 The figures can vary depending on, for example, whether the survey is talking about 'market share' (i.e. numbers of new computers sold with a particular OS) or 'installed base' (i.e. the number of computers currently in use and using a particular OS).

8 A Mozilla Foundation spokesman told silicon.com (October 2004) that they expected Firefox's share of the internet browser market to reach 10 per cent by the end of 2005. After its release in November 2004, ten million downloads were recorded in the first month alone.

7 Legal issues relating to construction collaboration technology

1 This chapter is an updated and expanded version of the author's contribution to Jones *et al.* 2003 (Wilkinson 2003b).

2 Developed by the Electronic Commerce Association, now e-centre (http://www.e-centre. org.uk).

3 Arguably, the situation in the United States is clearer. The Uniform Electronic Transactions Act (UETA), developed to recognise the validity of electronically formed contracts, records and signatures, has been enacted by many US states. Where UETA has not been adopted, the Electronic Signatures in Global and National Commerce Act (E-SIGN Act) also gives electronic contracts the same legal effect as hard copy forms. 'In other words, a contract cannot be denied validity just because it was created online or signed electronically' (Berning and Dively-Coyne 2000). It also allowed the retention of records in an electronic form, eliminating the need for hard copies of information.

4 For example, PD0008: 1999 (Code of Practice for Legal Admissibility and Evidential Weight of Information Stored Electronically), PD5000: 1999 (Electronic Documents and e-Commerce Transactions as legally admissible Evidence) and PD0010: 1997 (The Principles of Good Practice for Information Management).

5 As mentioned in Chapter 4, BIW and Cadweb provide ASP services conforming to BS7799.

6 Customers need to ensure that the terms and conditions of the End User Licence Agreement (EULA) with the ASP are appropriate and reasonable so that all members of their supply chain will be content to sign the EULA. If the EULA is not accepted by all supply chain members, gaps will arise in the chain of liabilities.

7 It may not be adequate for an individual to simply click-through an on-screen EULA; if a dispute arises, it may be necessary to ascertain the identity and the authority of the person to enter into that Agreement on behalf of the supplier. Accordingly, an ASP may insist on a copy of the EULA signed by an authorised signatory.

8 If the client in the building project is not familiar with such tools it could be argued that the professional team ought to raise the issue themselves rather than waiting until it is raised when the contractor comes on board.

9 The validity and enforceability of such clauses in an ASPs standard terms is controlled by the Unfair Contract Terms Act 1977. See also, for example, St Albans City and District Council v. International Computers Ltd (1996) 4 All ER 481.

10 However, in a May 2004 survey, PMP Research discovered that almost 40 per cent of organisations did not have SLAs with their IT departments.

11 American writer Alan Joch (2002) relates an instance where an extranet provider did not have sufficient computing resources to support a large US design project: 'The massive drawing files and flood of electronic communications overloaded the extranet to the point of shutting it down.' With the client threatening to hold the architect responsible for lost time, only a system upgrade by the extranet host got the project back on track. In this instance, a SLA would have limited the architect's liabilities for technical issues over which it had no direct control.

12 To put these figures in perspective, with 99.5 per cent availability, the amount of annual down-time would be about 43.8 hours (or about 3.6 hours per month); 99.8 per cent availability would allow 54 minutes downtime per month.

13 For example, for some weeks during June and July 2004, Asite's website warned that, due to 'essential maintenance', service interruptions might occur between 20.00 and midnight.

14 A collaboration solution can be in use around-the-clock, seven days a week. Monitoring its system in one typical 24-hour weekday period, BIW found the first user login was made at 04.13 am GMT, the last at 23.21 pm; user sessions were also recorded on Saturdays and Sundays and on public holidays. A project may have a multi-national team, or have stake-holders who travel to different time-zones, and limitations on system availability could impact on such users.

15 Key issues here can range from compliance with international rules on corporate gover-nance and public accounting (e.g. Sarbanes-Oxley, Basel II) to dealing with disputes alleg-ing libel, bullying, sexual harassment, etc. In June 2004, Silicon.com reported that two businesses (Perot Systems Europe and UBS) were contesting a request by an IT manager suing his former employers for £2 million over an alleged defamatory job reference; the cost of providing copies of email records dating back to 1999 was said to be £4.27 million, relating to searches on 70 servers, plus back-up tapes.

16 For example, Part VA of the Local Government Act 1972, the Environmental Regulations 1992, the Data Protection Act 1998, and the Local Government Act 2000.

17 Exemptions allow public authorities to withhold some or all of the information requested, but many will only apply where pressing public interest arguments can be made for with-holding it. Various absolute (e.g. it is personal information covered by the Data Protection Act or it is information that has been provided in confidence) and qualified exemptions apply (e.g. it prejudices a party's commercial interests).

18 Archiving policies should follow those for hard copy documents, and, according to the Construction Industry Council's liability briefing on e-business (2003): 'The general rule of thumb is to retain/archive key project documents for a period of no less than 17–20 years'.

8 Human aspects of collaboration technology

1 Many knowledge management system deployments have become unstuck because some users were unwilling to share their information and use the system, perhaps because companies failed to tie knowledge management to a business objective, meaning that it was viewed as just another IT initiative or simply another software application.

2 Responding to the silo-mentality, some AEC organisations have begun to use knowledge man-agement tools such as Union Square's Workspace to try to share enterprise and project-related information within their businesses.

3 The UK construction industry is very fragmented with the vast majority of industry businesses classed as small/medium-sized enterprises (SMEs). Among AEC professional services businesses, for example, 81 per cent have fewer than ten employees (Construction Industry Council 2004).

4 Concerns about the transparency of a system may make some users reluctant to comment publicly on somebody else's drawing, while the originator may not like critical comments on his or her work being publicly viewable.

5 Many central government departments and agencies have looked to deliver schemes via the Private Finance Initiative (PFI) or by Public Private Partnerships (PPPs), for example in the health sector, various NHS Trusts have been engaged in PFI through ProCure21 and LIFT schemes, and the UK Ministry of Defence has adopted prime contracting approaches for

some construction schemes. Some of the major water undertakings (e.g. United Utilities) have framework agreements with consortia of contractors, consultants and suppliers to deliver asset management plans (AMPs) worth hundreds of millions of pounds – airports operator BAA also uses framework contracts to ensure continuity of work for key supply chain partners.

6 Some prolific clients tested different systems simultaneously. For example, airport operator BAA used services from 4Projects, Bidcom and BIW before narrowing its choice to Bidcom and BIW (resold via Asite); retailer Asda used BIW and Sarcophagus before opting for the latter.

7 Materials can also be downloaded from the PIX Protocol website, www.pixprotocol.org. An online service – PIX Online (www.pixprotocol.com) – was added in November 2004, allowing project leaders and team members to input, store, manage and exchange their protocol information via the web.

8 When implementing systems for its customers, BIW, for example, offers different training to basic users, to company information co-ordinators, to company administrators, and to project information co-ordinators. Access, viewing and publishing tasks can be learned in a half-day, while more advanced administrative and system management skills may require additional modules. Nonetheless, up to two-and-a-half days of training is still significantly less than that involved in achieving proficiency in many general and professional IT applications, such as spreadsheets, CAD, etc.

9 For example, a December 2002 article in CADserver (www.cadserver.co.uk) said 4Projects would charge fees of £1,000 set-up and £1,000 per month for a £10 million project of 15 months duration. 4Projects Enterprise for an annual construction value of £200 million would be £9,500 to set-up and then £9,500 per month. The same £10 million project would cost £380 per month.

10 As far as some designers are concerned, the production of paper may simply be replaced by the conversion of native CAD files into other formats used by the collaboration system. While quicker than production and distribution of paper drawings, this can still seem quite laborious or time-consuming (some vendors have developed tools that automate the file conversion/creation and publication process), and designers might include a small allowance for such processes when agreeing timetables for production of design information.

11 This section incorporates material from the author's contribution to Preece et al. (Wilkinson 2003a).

12 Construction manager PCM, for example, set up Knowledge Online in early 2000, offering services from BIW, BuildOnline, Cadweb and Causeway Technologies, among others.

9 Benefits of using construction collaboration technologies

1 In 2004, researchers at Harvard University (including Professor Spiros Pollalis and Burcin Becerik) were gathering support from several collaboration technology vendors (mainly US-based, although some UK firms were also approached to participate) for a project intended to develop and validate a ROI model.

2 Some researchers have suggested more elaborate groupings of benefits. For example, Nitithamyong and Skibniewski (2004) identified six areas of potential performance improvement: strategic-related, schedule/time-related, cost-related, quality-related, risk-related and communication-related.

3 An ROI calculator for Autodesk Buzzsaw assumed its solution would save (1) US$1,150/month over an FTP server, (2) two hours of system maintenance and management time per project per month over an internally hosted website, (3) 30 minutes per project per month over an email-based system, and (4) US$1,500 on purchase and installation of a data protection firewall.

4 A collaboration tool can be used to collate detailed statistics showing how long particular processes take. For example, BIW staff monitored a commercial project in London and found that, of some 1300 RFIs, more than 70 per cent were completed in less than 14 days, while more than 50 per cent of almost 300 instructions were also completed in under a fortnight. Collected across successive projects, such statistics would help teams quantify the benefits of using the system.

5 Use of a single project repository also reduces the number of IT systems that can go wrong. The author learned of a project where a contractor was unable to use email for a month due to a virus attack. Fortunately, the contractor was able to use the project's collaboration platform as an alternative method of sharing information.

6 Arguably, much the same argument could be applied to drawings which are distributed via email (but which, of course, are not tracked or audited).

7 Such objections are similar to firms' concerns about partnering, and may have more to do with maintaining the status quo where they have been able to exploit the problems and inefficiencies of traditional fragmented and adversarial project delivery processes. Greater transparency and clear audit trails will clearly threaten such approaches; as stressed several times, successful collaboration is more about people and processes than it is about technologies.

8 For example, see the case study of the MOD prime contract undertaken at Andover North using BIW's system (IT Construction Best Practice (ITCBP) Programme 2003b).

9 See also Section 9.6.3.

10 A version of this case study is included in Sun and Howard 2004.

11 Based on case study published by ITCBP Programme (2003a).

12 Of course, if the technology is to be hosted by a member of the project team, then that team member may incur additional hosting costs, for example: the expense of a new server, a faster internet connection, new security measures, in-house IT department staff time, training, etc. (see Cohen 2001, p. 203).

10 Where next for construction collaboration technologies?

1 The IPO optimism of the dot.com boom evaporated when the bubble burst in 2000, and the ensuing scepticism took a long time to dissipate, but there were encouraging signs of a more positive, pragmatic approach to investment in technology companies during 2004 – witness the US flotations of Google and of ASP salesforce.com, for example.

2 Depending on what conditions (if any) are included in customer contracts about any change of technology provider ownership, room for manoeuvre may be limited.

3 The IFC initiative closely parallels STEP (STandard for the Exchange of Product model data), initiated by the International Standards Organisation in 1984, but is more specific to the AEC industry.

4 The NIST report defined interoperability as 'the ability to manage and communicate electronic product and project data between collaborating firms' and within individual companies' design, construction, maintenance, and business process systems' (p. ES-1).

5 For example, users within an architectural practice with an existing drawing management system wanted to be able to record the publication of drawings simultaneously to both their in-house SER system and to the BIW collaboration system they was using on a major project. BIW used XML to enable the integration.

6 As an indication of take-up, the same survey found that, of those using 3D applications for BIM, only 25 per cent – all of them structural engineers or from multi-disciplinary practices – used BIM for more than half of their production work.

7 For example, in December 2004 the suggested retail price for Autodesk Revit 7 was US$4,495.

8 Short for 'Wireless Fidelity'. This refers to the IEEE.802.11 industry standards used to create most WLANs.

9 Security can obviously be a major concern, as can the inability of users to roam across hotspots delivered by different service providers. One solution is extensible authentication protocol (EAP) SIM cards, used to verify a user's identity and authorise their connection to the WLAN, providing the subscriber with greater roaming coverage, ease-of-use and a single bill with his/her chosen service provider.

10 The basic standard for each technology is slightly different: IEE 802.16 for WiMAX and IEE 802.20 for MobileFi.

11 For example, South African-based Murray et al. Lai (2001) pointed out that the 'the use of web-based project sites could lead to increased efficiency and economy in the project

chain', but recommended that governments in sub-Saharan African countries needed to: 'maintain reliable telephone systems and/or encourage the setting up of cell phone and broad band networks' (p. 12).

12 Further information on the whole area of mobile applications for construction can be found at www.comitproject.org.uk

13 New technological developments may increase this reliability still further. Self-healing systems, hardware redundancy and load balancing across clustered servers may further reduce the number of staff needed in in-house IT departments.

Bibliography

Ahmed, S.M., Ahmad, I., Azhar, S. and Arunkumar, S. (2002) *Current State and Trends of E-Commerce in the Construction Industry: Analysis of a Questionnaire Survey*, Miami, FL: Florida International University.

Alshawi, M. and Ingirige, B. (2001) *Web-Enabled Project Management*, Salford: School of Construction and Property Management, University of Salford.

Autodesk (2003) *Opportunities and Unfinished Business: The Autodesk 2003 Survey of Issues and Trends in the UK Design Community – Focus on Building & Construction*, Farnborough: Autodesk.

—— (2004) *Autodesk Introduces Buzzsaw Server Edition*, press release, 11 February 2004.

Barbour Index (2001) *The Barbour Report 2001: Construction Product Information – Delivery Preferences and Trends: A Guide for Building Product Manufacturers*, Windsor: Barbour Index.

—— (2002) *The Barbour Report 2002: Exploring the Web as an Information Tool: A Practical Guide for Building Product Manufacturers*, Windsor: Barbour Index.

—— (2003) *The Barbour Report 2003: Influencing Clients: The Importance of the Client in Product Selection*, Windsor: Barbour Index.

Becerik, B. (2004a) *Suggestions for Improving Adoption of Online Collaboration and Project Management Technology*, 20th Annual Conference of Association of Researchers in Construction Management Proceedings, 1–3 September 2004, Edinburgh.

—— (2004b) A Review on Past, Present and Future of Web Based Project Management & Collaboration Tools and Their Adoption by the US AEC Industry, *International Journal of IT in Architecture, Engineering and Construction*, Vol. 2, Issue 3, October 2004, pp. 233–248.

Bennett, J. and Jayes, S. (1995) *Trusting the Team*, Reading: Centre for Strategic Studies in Construction, The University of Reading, with the partnering task force of the Reading Construction Forum.

Berning, P.W. and Diveley-Coyne, S. (2000) E-Commerce and the Construction Industry: The Revolution is Here, www.constructionweblinks.com. 2 October 2000. Online: Available <http://www.constructionweblinks.com/Resources/Industry_Reports_Newsletters/Oct_2_2000/e-commerce.htm> (accessed 11 May 2004).

Berning, P.W. and Flanagan, P. (2003) E-Commerce and the Construction Industry: User Viewpoints, New Concerns, Legal Updates on Project Web Sites, Online Bidding and Web-Based Purchasing, www.constructionweblinks.com. 22 December 2003. Online: Available <http://www.constructionweblinks.com/Resources/Industry_Reports_Newsletters/December_22_2002/e_commerce.htm> (accessed 11 May 2004).

Berning, P.W. and Ralls, J.W. (2002) Project Web Sites: How to Evaluate Them and How to Agree on Using Them, www.constructionweblinks.com. 13 May 2002. Online: Available <http://www.constructionweblinks.com/Resources/Industry_Reports_Newsletters/May_13_2002/project_websites.htm> (accessed 11 May 2004).

Birkby, G. and Nugent, J. (2002) The ASP with a Sting in its Tail, *Building*, 8 June, pp. 50–51.

Bjork, B. (2002) *The Impact of Electronic Document Management on Construction Information Management*, International Council for Research and Innovation in Building and Construction, CIB W78 Conference, Aarhus School of Architecture, 12–14 June 2002.

Breetzke, K. and Hawkins, M. (2003a) *An Introduction to Project Extranets and E-procurement*, London: RICS Construction Faculty.

—— (2003b) *Project Extranets and e-Procurement: User Perspectives*, London: RICS Construction Faculty.

BT Openworld (2002) *Building Companies Held Back by Poor Internet Plumbing*, press release, 30 July 2002.

Building Centre Trust (1999) *IT Usage in the Construction Team*, London: Building Centre Trust.

—— (2000a) *Extending Use of an Image Management System to SME Trade Contractors: Bluewater Retail Park*, ITCBP detailed case study no. 12, June 2000, London: Building Centre Trust.

—— (2000b) *Making the Connection – A Low Cost 'Extranet' for Construction Project Teams: Damond Lock Grabowski*, ITCBP detailed case study no. 20, September 2000, London: Building Centre Trust.

—— (2004) *PIX Protocol Guide and Toolkit*, London: Building Centre Trust.

Business Advantage Group (2002) *The Project Hosting Scene*. January. Online: Available <http://www.business-advantage.co.uk/Spaghetti/project_hosting.htm> (accessed 5 May 2004).

Business Vantage (2002) *Equal Partners: Customer and Supplier Alignment in Construction*, Windsor: Business Vantage.

—— (2004a) *Equal Partners: Customer and Supplier Alignment in Public Sector Construction*, Windsor: Business Vantage.

—— (2004b) *Equal Partners: Customer and Supplier Alignment in Private Sector Construction*, Windsor: Business Vantage.

Butler Group (2003) *Workgroup and Enterprise Collaboration – Reducing the Costs and Increasing the Value of Collaborative Working*, Hull: Butler Group.

Butler, M. (2003) 'Intellectual Property Rights', in D. Jones, D. Savage and R. Westgate (eds), *Partnering and Collaborative Working*, London: Informa, pp. 191–213.

CABE/Treasury (2000) *Better Public Buildings*, London: Department of Culture, Media & Sport.

Cain, C. (2001) *A Guide to Best Practice in Construction Procurement*, Watford: Construction Best Practice Programme.

—— (2002) 'Why Change Anything?' paper presented at *Chain Reaction: Technology Before Culture or Vice Versa?* conference, 29 November, Royal Institute of British Architects, London.

Champy, J. (2002) *X-Engineering the Corporation*, London: Hodder & Stoughton.

Chien, H.J. (2003) *Questionnaire Survey Results and Discussion*, Pontypridd, University of Glamorgan. Email, 20 July 2003.

Christian, M. (2003) 'The ASP Revolution is Dead: Long Live the ASP Guerrilla War!', ASPNews.com, 27 May. Online: Available <http://www.aspnews.com/trends/article.php/2212331> (accessed 27 April 2004).

Cochrane, P. (2004) 'Uncommon Sense: Are IT Departments Doomed?' Silicon.com. 7 May 2004. Online: Available <http://management.silicon.com/itpro/0,39024675,39120478,00.htm> (accessed 7 May 2004).

Cohen, J. (2000) *Communication and Design with the Internet*, New York: W.W. Norton.

Compagnia (2003) *Collaboration Software in the Construction Industry*, Ashford: Compagnia.

Confederation of Construction Clients (2000) *Charter Handbook*, London: CCC.

Construct IT (2003) *How to Manage e-Project Information*, Salford: Construct IT/ITCBP.

Construction Confederation (2001) *E-business Survey: Spring 2001*, London: Construction Confederation.

Construction Industry Board (1997) *Partnering the Team*, London: Thomas Telford.

Construction Industry Computing Association (2003) *Guidance on the Introduction and Use of Construction Extranets*, Cambridge: CICA.

Construction Industry Council (2003) *Liability Briefing on e-Business*, London: CIC. Online: Available <http://www.cic.org.uk/activities/eBus.shtml> (accessed 4 January 2005).

—— (2004) *Built Environment Professional Services Skills Survey 2003/2004*, London: CIC. Online: Available <http://www.cic.org.uk/activities/skill.shtml> (accessed 10 August 2004).

Construction IT Security Forum (CITSEC) (2004) *IT Security in Collaborative Environments*. Online: Available. <http://www.citsec.com/it-security-in-collaborative-environments.htm> (accessed 16 September 2004).

Construction Users Round Table (CURT) (2004) *Collaboration, Integrated Information and the Project Lifecycle in Building Design, Construction and Operation*, Cincinnati, OH: CURT.

Crane, A. and Saxon, R. (2003) 'The Future', in D. Jones, D. Savage and R. Westgate (eds), *Partnering and Collaborative Working*, London: Informa, pp. 51–62.

Crane, A. and Ward, D. (2003) 'The Story so Far', in D. Jones, D. Savage and R. Westgate (eds), *Partnering and Collaborative Working*, London: Informa, pp. 1–26.

Croser, J. (2003) 'Extranet Extras', *Architects' Journal*, 20 March, pp. 73–74.

Cyon Research (2004) *The Adobe Solution for AEC*, Bethesda: Cyon Research.

Davies, N. (2004) *CAD Managers' Survey Results: UK Architectural Engineering and Construction Industry*, London: Cadconsultancy. Online: Available <http://www.cadconsultancy.co.uk/downloads/Cadconsultancy-CADManagersSurveyResults.pdf> (accessed 1 November 2004).

Davis, M. (2003) 'Earning Interest on Knowledge Capital', *Butler Group Review*, June, pp. 21–22.

Department of Trade and Industry (2004) *Business in the Information Age: The International Benchmarking Study 2004*, London: DTI. Online: Available <http://www.dti.gov.uk/bestpractice/assets/ibs2004.pdf> (accessed 3 December 2004).

Duyshart, B. (1997) *The Digital Document: A Reference for Architects, Engineers and Design Professionals*, Oxford: Architectural Press.

Egan, J. (1998) *Rethinking Construction, Report of the Construction Task Force*, London: HMSO.

—— (2002) *Accelerating Change, Consultation Paper by Strategic Forum for Construction*, London: HMSO.

Evans, R., Haryott, R., Haste, N. and Jones, A. (1998) *The Long Term Costs of Owning and Using Buildings*, London: Royal Academy of Engineering.

Faithfull, M. (2003) 'Web of Ties', *Electrical and Mechanical Contractor*, March.

Fedeski, M. and Sidawi, B. (2002), *The Management of Internet Use in UK Architectural Practices*, Cardiff University. Online: Available <http://archmedia.yonsei.ac.kr/research/acadia2002_proceedings/acadia2002/6/S6_6.pdf> (accessed 13 October 2004).

Gates, B. (1996) *The Road Ahead*, New York: Viking Publishing.

Goodwin, P. (2001) *Effective Integration of IT in Construction: Final Report*, London: Building Centre Trust.

Green, S., Newcombe, R., Fernie, S. and Weller, S. (2004) *Learning across Business Sectors: Knowledge Sharing between Aerospace and Construction*, Reading: University of Reading/EPSRC/IMRC.

Hammer, M. (2001) *The Agenda*, New York: Random House.

Hammond Suddards Edge (2001) *The Role of Electronic Information in Construction Contracts*, BIW White Paper, London: BIW Technologies.

Hampton, J. (2001) 'Stung', *Construction Manager*, September, pp. 8–10.

Hansford, M. (2002) 'Virtual Togetherness', *New Civil Engineer*, 5 December, pp. 22–23.

Holti, R., Nicolini, D. and Smalley, M. (2000) *The Handbook of Supply Chain Management*, London: CIRIA.

Howard, R. (2002) *A Study of the Productivity Benefits of some Process Changes in the Building Industries of Denmark, Sweden and the UK*, Technical University of Denmark.

Institute of Directors (2002) *Software as a Service*, London: Institute of Directors.

IT Construction Best Practice Programme (2003a) *Adopting Internet Based Project Collaboration Software* (ITCBP case study 039), London: ITCBP. Online: Available <http://www.itconstructionforum.org.uk/uploadedfiles/039KIER.pdf> (accessed 31 August 2004).

—— (2003b) *Using a Project Extranet to Support Partnering in a Prime Contract* (ITCBP case study 041), London: ITCBP. Online: Available <http://www.itvconstructionforum.org.uk/uploadedFiles/041ANDOVER.pdf> (accessed 31 August 2004).

IT Construction Forum (2004) *Survey of IT in Construction: Use, Intentions and Aspirations*, London: IT Construction Forum.

Joch, A. (2002) 'E-documents Require due Diligence', *Architectural Record*, October. Online: Available <http://archrecord.construction.com/features/digital/archives/0210da-1.asp> (accessed 9 May 2004).

Kalay, Y.E. (1999) *The Future of CAAD: From Computer-Aided Design to Computer-aided Collaboration*, CAAD Futures '99 Conference, 7–8 June 1999, Atlanta, GA.

Kinns, J. (2000) *E-Construction: Management of Project Information*, Briefing sheet, London: Institution of Civil Engineers.

Lamont, Z. (2002) 'IT is Not the Answer', *Building*, 19 April, p. 33.

Latham, M. (1994) *Constructing the Team*, London: HMSO.

—— (2004) 'The Cynic's Bestiary', *Building*, 30 January, p. 20.

Leigh, J. (2004) 'Escrow Dividend', *Information Age*, March, p. 11.

Liker, J. (2003) *The Toyota Way: 14 Management Principles from the World's Greatest Manufacturer*, New York: McGraw Hill.

McBride, K. (2003) 'Clicks and Bricks: Practical Technology in Construction Contracts', *Plan Magazine*, October.

McCabe, L. (2003) 'Let's Separate the Two sides of the ASP Coin', *ASPnews.com*. Online: Available <http://www.aspnews.com/analysis/analyst_cols/article.php/1561691> (accessed 14 January 2005).

Magub, A. and Kajewski, S. (2003) *Identification of Skill, Knowledge and Abilities for the Use of the Internet for Information Sharing on Construction Projects*, Second International Conference on Construction in the 21st Century, 'Sustainability and Innovation in Management and Technology', 10–12 December 2003, Hong Kong.

Martin, J. (2003) *e-Procurement and Extranets in the UK Construction Industry*, FIG Working Week 2003, Paris, France, 13–17 April.

Murphy, L. (2001) 'Does the use of Construction Project Extranets Add Value to the Procurement Process?' unpublished dissertation for MSc in Facilities Management, London: South Bank University Faculty of the Built Environment.

Murray, M., Nkado, R. and Lai, A. (2001) *The Integrated Use of Information and Communication Technology in the Construction Industry*. Online: Available <http://buildnet.csir.co.za/constructitafrica/authors/Papers/w78-068.pdf> (accessed 28 April 2005).

National Audit Office (2001) *Modernising Construction: A Report by the Comptroller and Auditor General of the NAO*, London: HMSO.

National Institute of Standards and Technology (NIST) (2004) *Cost Analysis of Inadequate Interoperability in the US Capital Facilities Industry*, Gaithersburg, MD: NIST.

Nitithamyong, P. and Skibniewski, M.J. (2004) 'Web-based Project Management Systems: Survey of Success and Failure Factures', under review by *Building Research and Information Journal*.

O'Brien, W.J. (2000) 'Implementation Issues in Project Web Sites: A Practitioner's Viewpoint', *Journal of Management in Engineering*, May/June, pp. 34–39.

Ofcom (2004) *Oftel's Internet and Broadband Brief, April 2004*. Online: Available <http://www.ofcom.org.ukresearch/consumer_audience_research/int_bband_updt/may2004/> (accessed 11 June 2004).

Office of Government Commerce (2003a) *Achieving Excellence in Construction: Building on Success*, London: OGC.

—— (2003b) *Procurement Guide 05: The Integrated Project Team: Teamworking and Partnering*, London: OGC.

Orr, J. (2002) *Keys to Success in Web-Based Project Management: Lessons Learned from the Chicago Transit Authority Capital Improvement Program*, Bethesda, MD: Cyon Research.

PriceWaterhouseCoopers/DTI (2002) *Information Security Breaches Survey 2002*, London: PWC/DTI.

Sawyer, T. (2004) 'Web-based Collaboration: Online Management Tools Excel at Empowering Project Teams', *Engineering News Record*, 11 October 2004.

Schrage, M. (1990) *Shared Minds: The New Technologies of Collaboration*, New York: Random House.

Spekman, R.E. and Isabella, L.A. (2000) *Alliance Competence: Maximizing the Value of Your Partnerships*, New York: John Wiley.

Strategem/DTI (2003) *e-Business Sectoral Impact Assessment for General Building Contracting within the UK Construction Industry*, London: Department of Trade and Industry.

Strategic Forum for Construction (2003) *Integration Toolkit*, London: Strategic Forum for Construction. Online: Available <http://www.strategicforum.org.uk/sfctoolkit2/home/home.html> (accessed 7 May 2004).

Sulankivi, K., Lakka, A. and Luedke, M. (2002) *Project Management in the Concurrent Engineering Environment*, Tampere, Finland: VTT Building and Transport.

Sun, M. and Howard, R. (2004) *Introducing IT in Construction*, London: Spon Press.

Taylor, R. (1999) 'Computing Power on Tap', *Financial Times*, 1 November, p. 23.

Tuckman, B. (1965), 'Developmental Sequence in Small Groups', *Psychological Bulletin*, 63, 384–399.

Unger, S. (2001) *Why ASP Computing will Dominate the A/E/C Industry*, White Paper, October 2001. Alpharetta, GA: Constructware. Online: Available <http://www.constructware.com/Common/Downloads/ASP_vs_client_server_White_Paper_10–01.pdf> (accessed 11 June 2004).

Verheij, H. and Augenbroe, F. (2001) *A Survey and Ranking of Project Web Site Functionality*, Atlanta, GA: Georgia Institute of Technology.

White, E. (2001) 'Legal Aspects of Project Extranets', *Making Extranets Work Harder*, conference, 6 December 2001, Institution of Civil Engineers, London.

Whitton, D. (2004) *Extranet Clauses*, Email, 30 November 2004.

Wilkinson, P. (2003a) 'Business Development and Collaborative Working', in C. Preece, P. Smith and K. Moodley (eds), *Construction Business Development: Meeting new Challenges, Seeking Opportunities*, Oxford: Butterworth, pp. 167–179.

—— (2003b) 'The Role of Technology', in D. Jones, D. Savage and R. Westgate (eds), *Partnering and Collaborative Working*, London: Informa, pp. 27–50.

Womack, J.P., Jones, D.T. and Roos, D. (1990) *The Machine that Changed the World*, New York: Rawson/MacMillan.

Wood, C. (2001) 'ASPs', *PC Magazine*, 1 July. Online: Available <http://www.pcmag.com/article2/0%2C1759%2C115986%2C00.asp> (accessed 4 May 2004).

Index